FARADAY, MAXW
AND THE
ELECTROMAGN
FIELD

FARADAY, MAXWELL, AND THE ELECTROMAGNETIC FIELD

HOW TWO MEN REVOLUTIONIZED PHYSICS

NANCY FORBES AND
BASIL MAHON

59 John Glenn Drive
Amherst, New York 14228

Published 2014 by Prometheus Books

Cover image of Michael Faraday *(top)* used with permission from the Royal Institution, London, UK/The Bridgeman Art Library

Cover image of James Clerk Maxwell *(bottom)* courtesy of the Master and Fellows of Trinity College, Cambridge

Jacket design by Jacqueline Nasso Cooke

Inquiries should be addressed to
Prometheus Books
59 John Glenn Drive
Amherst, New York 14228
VOICE: 716–691–0133
FAX: 716–691–0137
WWW.PROMETHEUSBOOKS.COM

18 17 16 15 14 5 4

Library of Congress Cataloging-in-Publication Data Pending

ISBN 978-1-61614-942-0 (cloth)
ISBN 978-1-61614-943-7 (ebook)

Printed in the United States of America

FROM NANCY FORBES

To my parents,
Charles and Florence Forbes

FROM BASIL MAHON

To my children, Tim, Sara, and Danny;
my daughter-in-law, Helen;
and my granddaughter, Emily

CONTENTS

ACKNOWLEDGMENTS

The seeds for this book were sown over thirty years ago when one of us was a graduate student in physics and heard a lecture by Nobel laureate C. N. Yang about the role of symmetry in twentieth-century physics. Yang spoke of the birth of a pristine and heretofore unheard-of concept born from the simple instruments found in Michael Faraday's laboratory in the Royal Institution: field theory. It was the rise of this idea that gave way to most of the major developments in modern physics and became the basis for today's reigning theory of matter and forces, the Standard Model. Yet Faraday's work, by itself, could not have had the monumental effect on physics it had without the subsequent efforts by Scottish physicist James Clerk Maxwell to put Faraday's idea into mathematical language, which resulted in a full-fledged theory of the electromagnetic field. Yang, who made his own fundamental contributions to field theory, truly understood how radical and how profound this new concept was.

It was a story that was begging to be told. Many people lent a hand in getting this book written. Correspondence with Faraday scholars Frank James and Ryan Tweney helped to clarify aspects of Faraday's work. Talks with Maxwell biographer Francis Everitt also added to the narrative. Discussions with friends such as Tony Fainberg, Phil Schewe, Allan Blaer, Will Happer, Freeman Dyson, Mal Ruderman, Barbara and Jeffrey Mandula, Louise Marlowe, Robert and Linda Avila, and Simki Kuznick helped to fine-tune the story and make the physics more accessible. Thanks to Lee Bartrop for his carefully drawn diagrams throughout the text, as well as to John Bilsland for the diagrams of Maxwell's spinning cells model. Thanks also to Tom Haggarty at the Bridgeman Art Library for making available many of the images found in the photo insert. Trinity College Library at

Cambridge University also gave us permission to reproduce several Maxwell images. Our editor at Prometheus Books, Steven L. Mitchell, recognized the importance of our story early on and has handled our queries with patience and graciousness.

We are grateful for permission from the publishers of two of Basil Mahon's earlier books to include short extracts: to John Wiley and Sons for the diagram and accompanying text describing Maxwell's spinning cells model in *The Man Who Changed Everything: The Life of James Clerk Maxwell* (2003), and to the Institution of Engineering and Technology for the descriptions of the electromagnetic power vector and the curl and divergence operators in *Oliver Heaviside: Maverick Mastermind of Electricity* (2009).

CHRONOLOGY

PRINCIPAL EVENTS IN THE STORY OF THE ELECTROMAGNETIC FIELD

1600	William Gilbert publishes *De Magnete* and proposes that Earth acts as a giant magnet.
1687	Isaac Newton publishes *Principia Mathematica.*
1733	Charles du Fay distinguishes between vitreous and resinous electricity.
1745	Pieter van Musschenbroek and Ewald von Kleist independently invent the Leyden jar, a device that stores electricity.
1747	Benjamin Franklin puts forward the idea of positive and negative electric charge.
1750	John Michell demonstrates the inverse-square law of magnetism.
1766	Joseph Priestley demonstrates the inverse-square law of electricity.
1785	Charles Augustin Coulomb carries out precise experiments to confirm the inverse-square laws of electricity and magnetism.
1800	Alessandro Volta invents the voltaic pile, or battery, thereby making it possible to generate continuous electric currents.
1820	Hans Christian Oersted shows that an electric current in a wire deflects a compass needle.
1820	André Marie Ampère begins to formulate a combined theory of electricity and magnetism based on action at a distance.
1821	Michael Faraday discovers the principle of the electric motor.

1825	François Arago shows that a compass needle rotates when suspended over a spinning copper disc.
1831	Faraday begins to compile his *Experimental Researches in Electricity*. Faraday discovers electromagnetic induction and the principle of the dynamo. Faraday introduces the concept of lines of magnetic force.
1832–1833	Faraday discovers the basic laws of electrolysis.
1845	Faraday introduces the term *magnetic field*. Faraday shows that a strong magnetic field rotates the plane of polarization of light.
1846	Faraday gives his "Ray-vibrations" talk at the Royal Institution. Faraday discovers diamagnetism and shows that all substances have magnetic properties.
1855–1856	James Clerk Maxwell publishes his paper "On Faraday's Lines of Force."
1861–1862	Maxwell publishes his paper "On Physical Lines of Force."
1864–1865	Maxwell publishes his paper "A Dynamical Theory of the Electromagnetic Field."
1873	Maxwell publishes his *Treatise on Electricity and Magnetism*.
1882	Oliver Heaviside and Josiah Willard Gibbs independently propose vector analysis and vector notation.
1885	Heaviside summarizes Maxwell's theory of electromagnetism in the four now-famous "Maxwell's equations." John Henry Poynting and Heaviside independently derive the formula for energy flow in an electromagnetic field.
1887	Abraham Michelson and Edward Morley attempt to measure the aether drift but instead demonstrate that the speed of light is independent of the motion of the observer.

1888 Heinrich Hertz produces and detects electromagnetic waves in free space.

1892 Hendrik Lorentz publishes his theory of the (then-hypothetical) electron.

1897 Joseph John Thomson discovers the electron.

1900 Max Planck proposes the quantum to explain the black-body radiation spectrum.

1901 Guglielmo Marconi sends a telegraph signal across the Atlantic Ocean.

1905 Albert Einstein explains how quanta produce the photoelectric effect and predicts the photon.

Albert Einstein publishes his special theory of relativity.

INTRODUCTION

It is 1888. Picture a large, sparsely furnished room. It has stout wooden tables and workbenches—a laboratory of some kind—but there are no retorts, Bunsen burners, or flasks of brightly colored liquid. Instead, the room is stocked with curious metal devices that have strange names: Rhümkorff coils, Knochenhauer spirals, Wheatstone bridges.[1] Their purpose is to probe the ways of the mysterious invisible phenomenon—electricity.

The room has a single occupant, a young man, handsome, neatly suited, and dark-haired with a close-trimmed beard and moustache. He is deftly assembling some apparatus on one of the long wooden tables. At one end he has constructed a circuit that will produce electric sparks across a narrow air gap between two metal spheres connected to the ends of the wires in the circuit. Ordinarily air doesn't conduct electricity but, if the two spheres are close together and the voltage is high enough, a spark will appear to jump across the gap, although it is really a series of very rapid sparks that jump back and forth, or oscillate, between the spheres. To each sphere he has attached a metal rod connected to a rectangular metal plate—he has learned that this will alter the frequency of oscillation. He presses a key to activate the circuit, and vivid blue sparks crackle across the gap between the spheres.

So far, so good; his primary circuit works, as it had the day before and the day before that. He turns his attention to a separate part of the apparatus that he calls his detector—a simple loop of wire with a very small gap between its ends that he can adjust with a screw. He holds the detector close to the sparking primary circuit, and faint sparks appear across its own gap. This happens, he reasons, when waves of energy pass from the primary circuit to the detector.

All this is familiar ground to him, but the next steps are untried and will, he hopes, be decisive ones. Switching off the primary circuit for the moment, he props up a large zinc sheet in a vertical position at the far end of the table. Its purpose is to act as a reflector, like a mirror. He places the detector on the table between the primary circuit and the zinc reflecting sheet, closes the blinds, waits for his eyes to adjust to the darkness, and then switches on his primary circuit. Turning his back on the sparks scintillating between the spheres, he looks for tiny sparks between the terminals of his detector. They appear, faint but unmistakable. Now for the step that will, if successful, establish the result he is seeking. He looks to see if the brightness of the sparks varies as he moves the detector slowly away from the primary circuit toward the reflecting zinc sheet. Indeed, it does. The sparks diminish to nothing, then grow again to their brightest, and then the cycle repeats. He knows that when any kind of wave is reflected back toward its source, it forms a standing wave, which appears to vibrate in place, like a guitar string. Hence, waves are being produced by the primary circuit and reflected by the zinc sheet. This is exactly what he wanted to find. Heinrich Hertz, professor of experimental physics at the Technische Hochschule in Karlsruhe, has made one of the greatest experimental discoveries in the history of science: he has proved beyond doubt the existence of electromagnetic waves.

As Guglielmo Marconi and others were soon to show, the commercial value of Hertz's discovery was immense. But he had no notion of this, nor, indeed, of any practical application. What had captivated Hertz and set him on his quest was a beguiling but strange scientific idea—the brainchild of British experimentalist Michael Faraday in the 1830s that had been raised into a full mathematical theory by the young Scot James Clerk Maxwell three decades later. Their idea was so different from anything that had gone before that many of the leading men of the time dismissed it as a flight of fancy. Others were simply baffled; they did not know what to make of it. But to Hertz it was a beautiful idea that rang true. All it lacked was physical proof, and his quest was to supply experimental evidence that would put the matter beyond dispute.

From the time of Newton, leading scientists had believed that the universe was governed by mechanical laws: material objects held energy and inflicted forces. To them, the surrounding space was nothing more than a passive backdrop. The extraordinary idea put forward by Faraday and Maxwell was that space itself acted as a repository of energy and a transmitter of forces: it was home to something that pervades the physical world yet was inexplicable in Newtonian terms—the electromagnetic field.

Faraday's first notion of lines of force, much derided at the time, grew into Maxwell's sophisticated mathematical theory, which predicted that every time a magnet jiggled, or an electric current was turned on or off, a wave of electromagnetic energy would spread out into space like a ripple on a pond, changing the nature of space itself. Maxwell calculated the speed of the waves from the elementary properties of electricity and magnetism, and it turned out to be the very speed at which light had been measured. He surmised that visible light is just a small band in a vast spectrum of electromagnetic waves, all traveling at the same speed but with wavelengths that might range from nanometers to kilometers. All this remained just a theory with more skeptics than adherents until a quarter of a century later, when Hertz emphatically verified it by producing and detecting what we would now call shortwave radio waves in his laboratory. The door to previously unimaginable regions of scientific knowledge was opened.

It is almost impossible to overstate the scale of Faraday and Maxwell's achievement in bringing the concept of the electromagnetic field into human thought. It united electricity, magnetism, and light into a single, compact theory; changed our way of life by bringing us radio, television, radar, satellite navigation, and mobile phones; inspired Einstein's special theory of relativity; and introduced the idea of field equations, which became the standard form used by today's physicists to model what goes on in the vastness of space and inside atoms.

Faraday and Maxwell have attracted their share of biographers, and rightly so. Aside from their genius, both were admirable, generous-spirited men who conducted their science with infectious enthusiasm and exuded the kind of charm that made people feel better about themselves and the world in general. But perhaps even more compelling than their individual life stories is the way that the two men from totally different backgrounds—a self-taught son of a poor blacksmith and a Cambridge-educated son of a Scottish laird—were brought together by their curiosity about the physical world and their determination to find out how it works. Although they met only late in Faraday's life, they formed a tremendously strong bond—they were united by their willingness to challenge entrenched scientific customs and conventions. The theory of the electromagnetic field is their joint creation and has its own story, intertwined with theirs and with its own set of diverse supporting characters. There were, for example, the American rake Count Rumford, who was instrumental in founding the Royal Institution, which gave employment to the impecunious, young Faraday; the brilliant but vain Humphry Davy, who was Faraday's inspiring mentor; the maverick Oliver Heaviside, who summarized Maxwell's theory into the four famous "Maxwell's equations"; and the hardworking Oliver Lodge, who discovered waves along wires but found he had been comprehensively scooped by Hertz.

Welcome to the story of the electromagnetic field.

CHAPTER ONE

THE APPRENTICE

1791–1813

There are many places where one could begin this story. Perhaps the best is the rugged moorland on the windswept western side of the Pennines in the north of England. This was the kind of country depicted by Emily Brontë in *Wuthering Heights*, sparsely populated by hardy souls who eked out a living in a land where crops barely grew and sheep had to search hard to find sustenance. It was home to Michael Faraday's forbears.

The Faradays had joined a small sect of Christians that came to be called the Sandemanians, after Robert Sandeman, a Scot who had broken from the Presbyterian Church of Scotland and come to England in the mid-1700s. The Sandemanians worked hard, lived simply, and spoke plainly. Rejecting all the trappings of the established churches, they held to the simplest possible form of Christianity—their whole doctrine was contained in the epitaph on Sandeman's tombstone: "The bare death of Jesus Christ, without a thought or deed on the part of man is sufficient to present the chief of sinners spotless before God."[1] Though friendly to outsiders and tolerant of those with different views, they largely kept to their own society. This didn't stop them doing business: Faraday's father, James, was a blacksmith, and among his uncles were a weaver, a grocer, an innkeeper, and a tailor.

The life of Michael Faraday might have begun and passed quietly in the remote seclusion of rural Westmorland but for the pressure of wider events. In the mid to late 1700s, Britain had been fighting rival colonial powers at sea for many years, and it finally lost an expensive war against its own colonists in America. The cost of these ventures had taken its toll at home, and a looming revolution in

France held the prospect of a new cross-Channel war. Meanwhile, the Industrial Revolution was drawing people from the English countryside to the towns and cities: farmworkers left the land for the ironworks, potteries, and textile mills. Against this background, trade in Westmorland fell, but James Faraday, newly married, had his wife, Margaret, to support and wanted to start a family. In 1786, he decided to leave his ailing smithy in Outhgill and try his luck in London.

The Faradays settled in a poor area called Newington Butts, about a mile to the south of London Bridge. Two children, Elizabeth and Robert, soon arrived, and a third, Michael, was born in 1791. James may have picked up some farrier work from a nearby coaching inn called the Elephant and Castle (which now gives its name to the whole neighborhood), but business was slack, his hopes of prosperity came to nothing, and, to make things worse, his health began to fail. The dread of being committed to a debtors' prison or, worse still, to the workhouse must have been ever present. At times he had to accept charity, probably from fellow Sandemanians, to keep his family from starvation. But he did so with no bitterness or envy of those more fortunate. The children grew up in a lively and loving household, though a crowded one—Michael was soon followed by a young sister, called Margaret after their mother. The family moved again, first to Gilbert Street and then to rooms over a stable near Manchester Square, just off the newly named Oxford Street. Now home to fashionable department stores, this road had other associations in Faraday's time: it was formerly Tyburn Street, along which condemned men made their last journey to the gallows from Newgate Prison.

After a rudimentary education, thirteen-year-old Michael Faraday began work as an errand boy for George Ribeau, who ran a book and newspaper shop on nearby Blandford Street. He became a familiar figure in the neighborhood—a lively boy with "a load of brown curls and a packet of newspapers under his arm."[2] Ribeau was a French émigré with progressive views who took a warm interest in his young charges. He thought he had found a gem in Faraday and soon took him on as an apprentice bookbinder—a seven-year contract in those days. As far as we know, none of Ribeau's apprentices actually became a bookbinder, but his liberal regime allowed them the freedom to develop their talents in other directions: one became a comedian and another a professional singer.

Binding books hour after hour, day after day, was tedious work, but it required immense care and skillful hands—qualities that were to serve Faraday well later in life. What opened up the world for him, though, were the books themselves: picture books, adventure stories, novels, books on philosophy, and, most of all, anything about the physical world and men's attempts to find out how it worked. He became a seeker of truth. As he later recalled: "I was a very lively, imaginative person. I could believe in the Arabian Nights as easily as in the Encyclopaedia. But facts were important to me and saved me. I could trust a fact but always cross-examined an assertion."[3]

The more he read on any topic, the more he became aware of his own lack of education. Then, just when he needed it, he found a book that could have been written for him alone, *The Improvement of the Mind*, by the Reverend Isaac Watts. Faraday threw himself with all the vigor of youth into Watts's program for self-improvement. He tried to learn correct speech and told friends to put him right every time he made a grammatical slip in conversation. He took every opportunity to widen his knowledge and began to keep a "commonplace book," setting down facts, especially scientific ones, for future reference. Even in these simple notes he strove to follow Watts's advice always to use precise language and be guided by *observed* fact. And he took to heart another of Watts's instructions: one should "be not too hasty to erect general theories from a few particular observations, appearances or experiments."[4] One can see here the seeds of the scientific method Faraday later made his own: take imagination to its limits but draw no conclusions without solid experimental proof.

Yet another of Watts's suggestions was to supplement reading with the "living discourse of a wise, learned and well-qualified teacher,"[5] so Faraday borrowed a shilling from his brother and went to a lecture on electricity by John Tatum, a silversmith who had founded the City Philosophical Society. This was a kind of common man's Royal Society, which held weekly meetings attended by a motley collection of self-improvers, all eager to hear of the latest scientific discoveries.

Faraday soon joined the City Philosophical Society and became its most diligent student; at every lecture he took rough notes that he copied out in a fair hand at home. He enjoyed the boisterous company of his fellow students, several of whom became lifelong friends. The closest was Benjamin Abbott, a Quaker with a job in a mercantile house. He became Faraday's soul mate, someone who could help him acquire the social poise and the speaking and writing skills he lacked, and to whom he could pour out his innermost thoughts. Faraday loved music and later, when he could afford it, became a keen opera-goer. As with all his interests, he had to try things for himself—he took up the flute and enjoyed taking the bass part in choral singing.

He also made his own experiments, for example, he would use glass jars lined with metal foil to store static electricity, with which he could charge up household objects and administer mild shocks to himself and anyone else who wanted to join in. He was already beginning to think about how electricity worked and questioned the truth of an ostensibly authoritative article in the *Encyclopaedia Britannica*. Its author, James Tytler, had confidently propounded Benjamin Franklin's "one-fluid" theory, which ascribed positive charges to an excess of a mysterious electrical fluid and negative ones to a dearth. Most British scientists favored this theory, but Faraday's early preference was for the French "two-fluid" theory, in which one type of fluid gave rise to positive charges and the other to negative. Even there, he thought there was something amiss with the standard interpretation. The young upstart was right to doubt his elders and betters, but the problem turned out to be more difficult than anyone had imagined. It was many years before Faraday came close to explaining static electricity to his own satisfaction, and it was half a century before his own follower, Maxwell, put the last piece in the puzzle.

Just as Faraday thought he was beginning to come to grips with electricity, his plans for further work were knocked sideways by an astonishing discovery made in Italy. John Tatum had learned of the voltaic cell, or battery, invented ten years earlier by Alessandro Volta, and he described it to his audience at a City Philosophical Society meeting. Familiar devices for storing electricity, like the foil-lined glass jars Faraday had used, released all their charge in one burst, but the battery produced something hitherto undreamed of—a continuous flow of electricity. What was more, the new electric currents could be used in simple experiments to investigate the structure of matter. A vast new region of science was opening up. And it was easy to make a battery: in 1800, Volta had provided detailed instructions. Make a stack of metal plates, alternately copper and zinc, interleaved with layers of pasteboard dampened with salt water, and, amazingly, an electric force will be generated through the stack—the more plates, the greater the force. Connect the two end plates with a metal wire, and a continuous current flows. This was not all. Experimenters had found that if they fixed a wire to each end plate and dipped the two wire ends in a solution of a chemical compound, the electric force would cause the constituent parts of the compound to separate, with one part gathering at one wire's end and one at the other.

Simply hearing or reading of such things was never enough for Faraday. When assessing the work of others, he always had to repeat, and perhaps extend, their experiments. It became a lifelong habit—his way of establishing ownership over an idea. Just as he did countless times later in other settings, he set out to demonstrate this new phenomenon to his own satisfaction. When he had saved enough money to buy the materials, he made a battery from seven copper halfpennies and seven discs cut from a sheet of zinc, interleaved with pieces of paper soaked in salt water. He fixed a copper wire to each end plate, dipped the other ends of the wires in a solution of Epsom salts (magnesium sulfate), and watched.

> Both wires became covered in a very short time with bubbles of some gas, and a continued stream of very minute bubbles, appearing like small particles, ran through the solution from the negative wire. My proof that the sulphate was decomposed was, that in about two hours the clear solution became turpid: magnesia [magnesium oxide] was suspended in it.[6]

If there was one moment that confirmed the course of Faraday's working life, this was surely it. Nothing less than a career in science would do. He had already found the perfect instructor: Jane Marcet, who spoke to him through her remarkable book *Conversations on Chemistry*. Originally published anonymously in 1805, it became hugely popular, especially in America, where pirated versions abounded. In a manner reminiscent of Galileo in his *Dialogue on the Two Chief World Systems*, the author's alter ego, Mrs. B., instills in her pupils, diligent Emily and flighty Caroline, a scientist's curiosity about the physical world and a delight in discovering a little about it. They do experiments, using materials from around the house, and learn about heat and light, all the while being careful to draw conclusions only from observed fact. All this chimed precisely with Watts's guidance on life in general, so even before Faraday's experimental career started in earnest he had the procedure clear in his mind: explore; observe; experiment; eliminate sources of error; compare theory with experimental findings; keep thinking; and finally, draw whatever conclusions stand the test but even then be open to challenge—don't become a prisoner of your own ideas.

The daughter of a rich London merchant, Jane Marcet had married Alexander, a Swiss-born physician, and had come to share his passion for science. They took a well-informed interest in the latest developments, and their social circle included many of the leading scientists. One can picture their elegant dinners, at which prominent fellows of the Royal Society might gather to discuss Thomas Young's wave theory of light or Johann Ritter's discovery of ultraviolet radiation. This world was far removed from Faraday's. The rules of society were well described in the hymn "All Things Bright and Beautiful," written a few years later by Fanny Alexander. One verse runs:

> The rich man in his castle,
> The poor man at his gate,
> God made them high and lowly,
> And ordered their estate.

Blacksmiths' sons and bookbinders' apprentices did not, indeed should not, aspire to friendship and fellowship with people of higher rank. But science transcends class distinctions, and Jane Marcet had mentioned in the preface to her *Conversations* the two factors that were to transform Faraday's life—the Royal Institution and Humphry Davy.

Compared with the venerable Royal Society, the Royal Institution was an upstart establishment. Some of its fifty-eight founders were already prominent fellows of the Royal Society, and they intended the new Institution to complement rather than rival its older sibling. Its purpose, formally agreed to at the inaugural meeting on March 7, 1799, was to further "the Application of Science to the Common Purposes of Life" by means of "courses of Philosophical Lectures and Experiments." It was financed initially by subscription, and the prime mover among the founders was one of the most extraordinary characters in the history of science—an anglicized American named Benjamin Thompson who bore the title Count Rumford. Thompson had overlapping careers as fortune hunter, rake, philanderer, spy, military governor, inventor, park designer, scientist, and social reformer, and he can be described fairly as outstanding in all these roles. His title of count (of the Holy Roman Empire) had been bestowed by a grateful elector of Bavaria for transforming the Bavarian army from a rabble into a fit and efficient fighting force—he chose the Rumford part of the title from the town in New Hampshire where, in a part of his life little known to his British colleagues, he had abandoned his wife and daughter two decades earlier. Thanks in the main to Thompson's vision and drive, the Institution gained its Royal Seal in 1800 and set up business at 21 Albemarle Street in London, an address it still occupies today. Ever unpredictable, Thompson decamped to Paris after nearly ruining the fledgling organization with his over-ambitious plans for educational courses, but before going he saved the situation by a masterstroke that would make the Institution famous and fill its coffers. He recruited Humphry Davy, a cocksure twenty-three-year-old from the West Country with a marked Cornish accent, to run the chemical laboratory and act as assistant lecturer.

Davy, debonair and strikingly handsome, had already acquired notoriety from some widely reported exploits with mind-altering nitrous oxide, laughing gas, so the people from London's fashionable elite who bought expensive tickets for his early lectures were probably drawn as much by the hint of scandal as by scientific interest. They were not disappointed—Davy amply satisfied their wish for excitement. But there was more: he enlightened and inspired them at the same time by presenting the wonders of science with romantic eloquence, passion, and pyrotechnical demonstrations. Word spread, and even bigger crowds came to the next course of lectures. Albemarle Street had to be made London's first one-way thoroughfare to prevent it from being blocked by carriages. When he was ill in 1807, there were so many inquiries about his health that the Royal Institution posted hourly notices outside its headquarters. Nothing like Davy's popularity had been seen before. Half his audience were women—many of them young, enraptured by the dashing young man. One wrote "those eyes were made for something besides poring over crucibles."[7]

Davy was not only a showman. His groundbreaking work in the Royal Institution laboratory put him in the front rank of scientists. He discovered the elements potassium and sodium by sending an electric current through molten compounds in order to separate their components. The method was the same in principle as that by which Faraday managed to decompose Epsom salts in Ribeau's workshop, but rather larger in scale: Faraday's battery, made from halfpennies and home-cut zinc discs, had seven voltaic cells while Davy's had two thousand. At the end of another series of experiments, Davy concluded that chlorine, the green gas given off when hydrochloric acid (then called muriatic acid) reacted with manganese dioxide (then called pyrolusite) did not contain oxygen, as was generally believed, but was an element in its own right. This was close to heresy according to followers of the great French chemist Antoine Laurent Lavoisier, who had assured everyone that oxygen was a necessary element in all acids, and it was some years before Davy's correct view prevailed.

Neither Faraday nor his friends at the City Philosophical Society could afford tickets to the lectures, but they read and heard of the great man's achievements. The knowledge that Davy was making scientific history only a few streets away was both thrilling and frustrating. Faraday felt the buzz, but it was like hearing the sounds of a party going on behind closed doors. Then one day, to his joy, he was given a ticket to a course of four lectures by Davy. It was his unfailing diligence at Society meetings that had brought this unexpected dividend. His carefully written and illustrated notes of the lectures were so beautifully done that Ribeau used to show them off to customers at the shop. One of them, Mr. Dance, who happened to be a member of the Royal Institution, was impressed beyond words by Faraday's work and straightaway presented him with the ticket.

Faraday arrived early for each lecture to be sure of getting the best seat in the gallery and sat with rapt attention as Davy put on a tremendous show, not only demonstrating his groundbreaking experiments but also conveying to the audience the joy of it all. As was his habit, Faraday took thorough notes of all that went on and wrote them up, with his careful penmanship, at home. Wonderful though Davy's demonstrations had been, Faraday felt the need to repeat the experiments and see the results himself. For him, this was the only way to understand what was really going on in the physical world. He did what he could, using makeshift equipment gathered from home or the shop, and dreamed that one day he might have a laboratory of his own.

With the end of his apprenticeship approaching, Faraday wrote to Sir Joseph Banks, president of the Royal Society, asking for a scientific post, no matter how menial, but had no reply. Checking with the porter at the Society's headquarters in Carlton Terrace, he discovered that Banks, or perhaps his secretary, had annotated his letter with the comment "No answer required."[8]

The apprenticeship over, he took a job as a bookbinder with Henri de la Roche, another French émigré. Work took all the hours of the day, and the outlook was bleak indeed. But Sandemanians didn't sulk, they just got on with the job; de la Roche was so impressed with Faraday that after only a few months, he offered to make him heir to the business. This was not what Faraday wanted. In despondency, he wrote to a friend:

> As long as I stop in my present situation (I see no chance of getting out of it just yet), I must resign philosophy entirely to those who are more fortunate in the possession of time and means.[9]

Just when things looked blackest, he had a request for help from, of all people, Humphry Davy.

Faraday's white knight, Mr. Dance, had come to his aid again. When Davy was temporarily blinded by an explosion in the laboratory and needed an assistant, Dance had said he knew just the man for the job. Faraday took leave from work and spent a few euphoric days in his hero's presence. It was soon over, and the vanished glimpse of a brighter world made the work routine more mind-numbing than ever. Nothing ventured, nothing gained: he carefully bound his notes of the lectures and sent them to Davy, asking if there was any chance of a permanent position at the Royal Institution. A few days later, on Christmas Eve 1812, he was overjoyed to get a reply.

> Sir,
>
> I am far from displeased with the proof you have given me of your confidence, and which displays great zeal, power of memory, and attention. I am obliged to go out of town till the end of January: I will then see you at any time you wish.
>
> It would gratify me to be of service to you. I wish it may be in my power,
>
> I am, Sir,
> Your obedient humble servant,
> H. Davy

Faraday kept the letter all his life, but it didn't bring an immediate reward. At the interview, Davy said that he would like to employ Faraday, but all the posts at the Institution were filled and none was likely to become vacant soon. Despondency returned, but not for long. One evening, a carriage pulled up outside the Faradays' house and a footman delivered a message. Would Mr. Faraday kindly visit Mr. Davy in the morning?

Greeting his eager visitor, Davy explained that he had had to sack his bottle washer for fighting, and he offered Faraday the post—with a worldly warning. Many years later, Faraday recounted to a friend what had happened:

> At the same time that he thus gratified my desires as to scientific employment, he still advised me not to give up the prospects I had before me, telling me that Science was a harsh mistress and in, a pecuniary point of view, but poorly rewarding those who devoted themselves to her service. He smiled at my notion of the superior moral feelings of philosophical men and said that he would leave me to the experience of a few years to set me Right on that matter.[10]

Sound advice, but, as Davy had probably surmised, there was no chance of its being accepted. Faraday began his life in science. The post was the lowliest in the Institution, informally designated "fag and scrub." There were not only bottles to wash but floors to sweep and fireplaces to clean. But it didn't take Davy long to recognize his new employee's talents. He first set Faraday to extracting sugar from beetroot, and soon the two were working together, braving the hazards of nitrogen trichloride, the treacherous compound that had once nearly blinded Davy and regularly blew its containing tubes and basins to pieces. For Faraday it was a second apprenticeship, this time in his proper vocation.

Davy was now the greatest man of science in Europe. Not only a fellow of the Royal Society, he had also been awarded the Napoleon Prize[11] by the Institut de France at a time when Britain and France were fighting a painful war. Whatever he turned his hand to seemed to make headline news everywhere. He was knighted by the king in 1812, and the same year, he married Jane Apreece, a rich young widow who had beguiled top-drawer society with her beauty, elegance, and wit. He had joined the haut monde, wining and dining with the privileged and the successful. The Davys had wide cultural interests and a special fondness for writers. She was a friend (and distant cousin) of Sir Walter Scott, and he loved poetry—he counted Robert Southey and Samuel Taylor Coleridge among his particular friends and often broke into poetic imagery himself. Coleridge said the reason he came to Davy's lectures was to refresh his stock of metaphors.

Now that he was a man of independent means, it was natural for Davy to want to travel to Europe—to visit the great cultural centers, meet the top scientists he had been corresponding with for years, and collect his Napoleon Prize in person. It was also natural for some of his countrymen to label him a traitor for even harboring such thoughts—France was Britain's bitter enemy and Napoleon was the devil in human form. Davy took a broader view. In a letter to a friend, he wrote:

> Some people say I ought not accept the prize; and there have been foolish paragraphs in the papers to that effect; but if two countries or governments are at war, the men of science are not. That would indeed be a civil war of the worst description; we should rather, through the instrumentality of men of science, soften the asperities of national hostility.[12]

Napoleon granted the necessary special passports, and plans were made for a party of five to depart from Plymouth: Davy, his valet, his wife, her maid, and Faraday as scientific assistant. A few months earlier, Faraday had feared he might be a bookbinder all his life. Now he was soon to embark on a Grand Tour of the kind normally reserved for sons of the aristocracy, and in the company of someone he admired beyond measure. But life with Davy was rarely smooth. When his valet backed out shortly before sailing, Davy asked Faraday to double up as temporary valet—just until Paris, where he would find a proper servant. This caused Faraday to think twice about going; he was already nervous at leaving his familiar surroundings, and now there was the humiliation of having to clean another man's boots. But what an opportunity would be lost! Pride was set aside, and in October 1813 he took his seat on the coach for Plymouth.

CHAPTER TWO

CHEMISTRY

1813–1820

It was the first time Faraday had traveled beyond the outskirts of London. New sights and sounds were all around. He sat atop the coach all the way to Plymouth, taking everything in, and, when the ship set sail and the Davys retired to their cabin, he stayed on deck, wrapped in a blanket, wide-eyed at his first view of the sea.

His first impressions of France could not have been worse. After docking at Morlaix in Brittany at the end of a rough two-day voyage, the English party had to wait hours for a pompous official to arrive and supervise a ritual of questioning and searching before they were allowed to set foot on land. Making their way in the dark along muddy tracks, they reached the hotel well past midnight to find all the hallways and corridors occupied by beggars warming themselves and looking for scraps of food, while chickens, pigs, and horses did the same.

Only as they approached Paris did they see signs of Napoleon's new France. The roads improved and squalor gave way to grandeur. At the Hotel des Princes, Faraday dealt with the hotel staff to make sure the Davys had everything they needed and took his first stroll along the boulevards. He had not walked far before the cry "*Anglais*" rose up and he was jostled and spat upon. His clothes had given him away. As soon as he could, he bought a new suit in the French style, but he still felt a chilling isolation; he wrote in his journal: "I know nothing of the language or of a single human being here, added to which the people are enemies and they are vain."[1]

Any self-pity was private and short-lived. To understand this strange, new world he had to join it, so he set himself to learning

French. Once he could exchange a few words, he began to like the French people a little better, finding them "communicative, brisk, intelligent and attentive,"[2] although over concerned with appearance and downright predatory when it came to monetary dealings. Ever the diligent and perceptive observer, he wandered Paris's sharply cobbled streets until his feet hurt, exploring halls, churches, gardens, monuments, and galleries, making meticulous notes in his journal. All heady stuff for a twenty-one-year-old newly released from a wearisome trade in London, but a black cloud hung overhead. He was a valet.

The work itself was not the problem. Davy, not being a born aristocrat, didn't need or want help dressing or shaving himself. What Faraday hated was the indignity, the humiliation, of being accorded the status of servant. As the weeks passed, he waited in vain for Davy to keep his promise to find a proper valet in Paris, but candidate after candidate was turned down as unsuitable. Even this might have been borne with a shrug of the shoulders but for Lady Davy. Faraday wrote of her to his friend Ben Abbott: "She is proud and haughty to an excessive degree and delights in making her inferiors feel her power."[3] Jane Davy understood nothing of her husband's work, and in her eyes Faraday was just a vassal with pretensions beyond his station.

But Faraday was nobody's servant when he joined Davy in lively sessions with top French scientists. Among them were Joseph Louis Gay-Lussac, with whom Davy had a (so far) friendly rivalry, and André Marie Ampère, with whom Faraday was to forge a similar (but longer-lasting) relationship. Davy had brought from England a portable laboratory stocked with enough materials to cause a fair-sized explosion. One wonders how he got it past the port officials at Morlaix, but he now used it freely in his hotel rooms, laying on demonstrations for guests. One day, Ampère and some colleagues brought in a small box filled with shiny, dark-grey flakes—they called it "substance X." It was actually an unintended by-product from a gunpowder factory, but all Ampère told Davy was that Gay-Lussac and others had tried, and so far failed, to identify its chemical composition. Davy went to work at once, heating a few of the flakes in a tube and producing for his guests a startling display of deep-purple gas. After making test after test over the next few days, using techniques that only he knew, he concluded that the mysterious substance must be a completely new element and called it "iodine," after the Greek word for *purple*.

He cautiously compared notes with Gay-Lussac, not wanting to give too much away. France's top chemist had already named the new substance iode and identified many of its properties but, after trying every test he could think of, was still not sure whether it was an element or a compound. So much for self-vaunting French scientists! Davy boldly staked a claim for himself and Britain by dashing off an announcement of his discovery to the Royal Society in London. Gay-Lussac was livid and claimed the credit for himself and France. Though they may have distanced themselves from the war, the men of science were not immune to chauvinism. Poor Ampère was denounced by his countrymen for giving Davy the opportunity to poach on Gay-Lussac's territory.

Faraday had no doubts on the matter. Describing Davy's brilliant work, he wrote to Abbott:

> The discovery of these bodies contradicts many parts of Gay-Lussac's paper on iodine, which has been much vaunted in these parts. The French chemists were not aware of the importance of the subject until it was shown to them, and now they are in haste to reap all the honors attached to it; but their haste opposes their aim. They reason theoretically, without demonstrating experimentally, and errors are the result.[4]

Little did he know that this last theme would come to dominate much of his later research on electricity and magnetism.

Having stirred up Paris, Davy and his party traveled on to Lyons, Montpellier, Aix, and Nice, then over the Maritime Alps in the middle of a harsh winter to Turin and Genoa, from where they took a ship to Florence. During a storm-tossed voyage, Lady Davy was seasick and couldn't speak for some time. Faraday later remarked to Abbott that her silence was well worth the risk to their lives. In Florence, Davy borrowed the duke of Tuscany's huge magnifying lens to burn diamonds by focusing the sun's rays on them. The diamonds were encased in a small glass vessel containing only oxygen and, when carbon dioxide gas was produced, Davy concluded that diamond must be a form of pure carbon, like soot, charcoal, and graphite.

Florence gave Davy and Faraday the chance to marvel at Galileo's instruments, including the famous telescope, in the Museo di Storia Naturale; and their next destination, Rome, offered the wonders of St. Peter's and the residual grandeur of the Roman Empire. What had happened to the Romans? Faraday wrote to Abbott:

> The civilization of Italy seems to have hastened with backward steps in latter [*sic*] years, and at present there is found there only a degenerate idle people, making no effort to support the glory that their ancestors left them.[5]

From Rome to Naples, then north again to Milan where Faraday and Davy met sixty-nine-year-old Alessandro Volta. Here was at least one living Italian whom Faraday admired. Davy recalled his own impressions:

> His conversation was not brilliant; his views rather limited, but marking great ingenuity. His manners were perfectly simple. . . . Indeed I can say generally of the Italian savants, that, though none of them had much dignity or grace of manner, they were all free of affectation.[6]

Then it was back over the Alps to Geneva, where Davy was keen to meet Gaspard de la Rive, with whom he had corresponded for years, and his son Auguste. The Davy party stayed at the de la Rives' lakeside villa for three months. Also in Geneva were Faraday's inspirational instructor Jane Marcet and her husband Alexander, probably visiting his Swiss relations. They asked the Davys and Faraday to dinner, but when the party arrived, Lady Davy—to the huge embarrassment of all but herself—ordered Faraday to take his supper with the servants. Alexander did his best to rescue the situation. As the ladies withdrew after dinner, he said "And now, my dear Sirs, let us go and join Mr. Faraday in the kitchen."[7]

Davy had, by now, made a remarkable impression on many of the scientists in Europe. So had Faraday. One of the young men they had met in both Paris and Geneva later wrote:

> His laboratory assistant, long before he had won his great celebrity by his works, had by his modesty, his amiability, and his intelligence, gained most devoted friends at Paris, at Geneva, at Montpellier. . . . Faraday has left memories equally charged with an undying sympathy which his master could never have inspired. We admired Davy, we loved Faraday.[8]

The tour went on. Back in Rome, Faraday was appalled, yet fascinated, to witness the way that apparently devout crowds filled the great churches to hear elaborately sung masses presided over by the pope and then straightaway launched into the licentious, week-long winter carnival. He longed for home, and to his joy the wish was fulfilled. Davy's plans to travel on to Greece and Turkey were knocked awry when plague broke out. There was a further impediment besides. From our distant viewpoint it seems extraordinary that Faraday and the Davys were able to roam Europe with scarcely a thought of the political situation. During their travels, Napoleon had been defeated by a coalition of most of the other European countries and exiled to Elba. Now he had escaped back to France and millions of his rejuvenated countrymen had rallied to his cause. Along with the British and Prussians, the Italians felt menaced, and troops gathered in the streets.

Faraday summed up his own lifelong views on the struggles of nations very nicely in his journal:

> I heard for news Bonaparte was again at liberty. Being no politician, I did not trouble myself much about it, though I suppose it will have a strong influence on the affairs of Europe.[9]

It is possible that Davy had yet another reason for changing his plans; perhaps the prospect of unrelieved proximity to his wife was becoming too much to bear. They took the quickest safe route, through Germany and Holland, to Ostend, and sailed for England. Faraday was overjoyed. He wrote to his mother from Brussels to say he would be home in three days, adding the postscript: "Tis the shortest and (to me) the sweetest letter I ever wrote you."[10]

After eighteen months away, the boy came back a man. He had seen Versailles, the Louvre, St. Peter's, the Coliseum, and Vesuvius, and he had crossed the Alps three times. He had mixed with Europe's elite scientists and struck up lasting friendships. He had endured Lady Davy's barbs and learned at length how to treat them with indifference. Most of all, he had been her husband's close companion, absorbing Davy's insight into the nature of chemistry and sharing in the process of scientific discovery with all its hard graft, false trails, doubts, and disappointments, along with the blissful moments of inspiration and exultation.

The Grand Tour had been a rite of passage for generations of young English aristocrats—the culmination of a privileged education that may well have included Eton and Oxford or Cambridge. A life-enhancing experience, though for some the appeal of Europe's finest art, music, and society was more than matched by the allure of Europe's finest fleshpots and gambling houses. Now, by a remarkable combination of luck and his own efforts, Faraday the blacksmith's son had trodden the same golden path. It was a rite of passage for him, too: he had greatly broadened his horizons and acquired much of the polish expected of a formally educated, young, English gentleman. It had opened his eyes to the world, including parts of it, such as fashionable society, that he wanted nothing to do with. Now he had seen far beyond the borders of his simple, hardworking mode of life and learned something of the ways of the rich and powerful. He could face them on equal terms as a man of the world.

Back at the Royal Institution, Faraday was given a modest pay rise and a curious job title, "Assistant and Superintendent of the Apparatus and Mineralogical Collection." "Fag and scrub" days were over, and valet duties were out of the question, but he still acted as Davy's amanuensis, bringing order to the great man's flamboyant but somewhat chaotic working life, writing up research notes, and keeping experiments going while Davy attended to his many outside interests, both professional and social. This duty was not light—as the months passed, Davy came to rely on him more and more—but for Faraday it was a labor of love, quietly performed in the shadow of Davy's charisma and stellar reputation. He copied out all his mentor's carelessly scrawled research notes in his own beautifully even hand, asking only that he be able to keep the originals, which he bound in special quarto volumes.

Faraday's own life was, by contrast, a model of organization. Indeed, it had to be, so that everything could be fitted in. Monday and Thursday evenings were for reading and other self-improving pursuits; on Wednesday evenings, he went to City Philosophical Society lectures, sometimes giving them himself; and Saturdays were always spent with his mother, leaving Tuesday and Friday evenings for keeping company with Ben Abbott and other friends. He loved it all.

In 1815, Faraday helped Davy develop the miner's safety lamp. Coal miners needed light to work, and hundreds had died in explosions ignited by open flames. In Davy's lamp, the flame couldn't pass through the fine, metal mesh that surrounded it; as long as the lamp was kept in good condition, there would be no explosion. Davy became the miner's hero, even though the lamp actually led to more deaths because it encouraged mine owners to reopen mines that had previously been closed for safety reasons.

Delighted with his protégé's progress, Davy gave Faraday projects that allowed him to publish under his own name. The name Michael Faraday began to appear in the journals, though his early papers gave little indication of what was to follow. They were short and plain, and they dealt with subjects unlikely to raise any pulses; for instance, the first was "The Analysis of Caustic Lime of Tuscany." But to Faraday these papers were precious: they signaled his accession to the ranks of practicing scientists. The thought that he might become a great discoverer didn't enter his head, but he was determined to do all he could to bring credit to his new profession. Part of his job was to assist his immediate boss, William Brande, with laboratory work and lectures. Where Davy dazzled in the lecture hall, Brande barely glowed, but he was a consummate professional and Faraday learned from him. Audiences at Royal Institution lectures were sparse now that Brande had taken that role from Davy, but Faraday had already formed views on the art of lecturing. As he had put it in one of his letters to Abbott: "the generality of mankind cannot accompany us one short hour unless the path is strewn with flowers."[11] Meanwhile, he practiced his own lecturing skills at the City Philosophical Society and attended evening classes in elocution and oratory, from which he filled 133 pages in a notebook. He could scarcely afford the fee for the course, but the time and money were well spent; he later became a master of timing and delivery who could hold an audience spellbound.

For now, he continued to build a reputation as a skillful and reliable chemist. Manufacturers and food producers had begun to recognize the need for precise chemical analyses, but there were very few people in the country able to carry them out—British universities didn't begin to teach practical chemistry until many years later. So Faraday, in his splendidly equipped laboratory, found himself in great demand from commercial companies and government departments. For example, he measured the water content in consignments of sodium nitrate supplied to gunpowder manufacturers, analyzed the gases emitted by aging eggs, and tested methods of drying various kinds of meat and fish to be used as food for sailors—a service performed for the Admiralty. Such business brought in much-needed income to the Royal Institution.

So did acting as expert witnesses in court cases. Faraday's senior colleagues generally took on the legal work, but in 1820 he was hired by a group of insurance companies to help defend their case, only to find that both Davy and Brande had been hired by the other party, a sugar-refining company. The insurers had refused to pay out after a fire, claiming that the refiner had invalidated the policy by using oil during refining. Faraday gave compelling evidence on the flammable nature of the oil but lost the case; the court somehow decided that the insurers should pay because the refiner had no intention to defraud. Perhaps the case turned on a fine point of law but, whatever the legal niceties, pocketing three fees from one case was good business for the Royal Institution.

On one occasion, Faraday let his standards slip. Davy, ever combative, published a paper on phosphorus compounds in which he challenged the findings of the Swedish chemist Jöns Jacob Berzelius. Included in the paper were some results from Faraday's work; Berzelius checked Faraday's results, found errors, and let fly:

> If M. Davy would be so kind as to take the pains of repeating these experiments himself he should be convinced of the fact that when it comes to exact analysis, one should never entrust them into the care of another person; and this is above all a necessary rule to observe when it comes to refuting the works of other chemists who have not shown themselves ignorant of the art of making exact experiments.[12]

This was utter humiliation, and a lesson. Never again did Faraday publish anything before doing his utmost to eliminate all possible sources of error.

Although *chemistry* is the word we use for the kind of work that Davy, Brande, and Faraday were doing, they didn't think of themselves as specialists but simply as men of science, or natural philosophers. They worked at chemistry because that is where the frontier of science was. Progress lay in discovering more about the composition of substances, and how they reacted when mixed or subjected to an electric current. Davy himself had broken fresh ground by isolating seven new elements. Scientists, as ever, were driven by a thirst for knowledge, and there was further motivation from industry. Manufacturers wanted to take advantage of the latest findings in chemistry—to make new products or to make old ones more economically—and were prepared to pay for research that could give them an edge: Faraday found himself visiting ironworks and being called in to try to improve the quality of steel used for surgical instruments. All in all, the way ahead seemed clear: more of the same was a thoroughly satisfactory strategy for anyone with intent to push back the frontier of science. But nobody could see what lay just over the horizon.

By the time of his twenty-ninth birthday in September 1820, Michael Faraday had established himself in the middle ranks of

British scientists. He was a first-rate chemical analyst—set, it seemed, for an honorable, if unspectacular, career as a stalwart of the Royal Institution. Nothing he had done so far seemed to signal momentous feats to come. Yet everything he had done to date turned out to be the perfect preparation for some of the greatest scientific achievements of all time. All of his faculties of observation, exploration, imagination, and contemplation, together with his experimental skill, meticulous record keeping, and sheer determination, would be tested to the full and not found wanting.

His call to arms came on October 1, 1820. Sir Humphry Davy arrived at the Royal Institution with some astonishing news from Denmark. Hans Christian Oersted had put a magnetic compass near a current-carrying electric wire and had seen the needle move to a position at right angles to the wire. In the twenty years since Volta gave them electric currents, scientists had been scrabbling in the undergrowth for scraps of knowledge while a discovery of the first magnitude lay on the path at their feet. The spirit of exploration was strong, so why had nobody else thought of placing a compass near an electric circuit to see if anything would happen? Strange though it seems to us, none but a few scientists thought there could be any connection between the forces of electricity and magnetism, and these few were regarded by the others as airy-fairy metaphysicians. The majority held firmly to what is generally called the Newtonian model, though Newton himself probably would have disowned some of it: material bodies inflicted forces by acting on one another instantaneously at a distance along straight lines. This model made no attempt to explain how one of nature's forces, like electricity, could interact with another, like magnetism, but, as we'll see, it had such a hold on scientific opinion that it clung on for many decades—even in the face of mounting evidence to the contrary, first from Oersted and then from Faraday's work and, later, Maxwell's.

Shocked and fascinated by the news of Oersted's discovery, Davy and Faraday naturally began to experiment with currents and magnets. And before long, Faraday would be combing the Royal Institution library, and other libraries, to see what could be gleaned from the history of electricity and magnetism.

CHAPTER THREE

HISTORY

1600–1820

Since ancient times, electricity and magnetism had been obscured by a fog of superstition, mysticism, and quackery. The man who began to dispel the fog was William Gilbert. Born in 1544 in Colchester, he trained as a physician and became a very good one, rising to be president of the College of Physicians and personal doctor to Queen Elizabeth. But we have still have more reason than his patients did to be grateful to him. He was the first to study electricity and magnetism experimentally, and his careful observation and scientific reasoning cleared the way for those who took up the work later.

Why did a suspended magnetic needle always align itself north–south? Why did amber attract pieces of paper and fluff after being rubbed with fur? Fascinated by such questions, Gilbert looked for enlightenment to works of scholarship, both ancient and contemporary, but found nothing that shed any light on the subject. In his book *De Magnete*, published in 1600, he reports:

> Many modern authors have written about amber and jet attracting chaff and other facts unknown to the generality: with the results of their labors booksellers' shops are crammed full. Our generation has produced many volumes about recondite, abstruse and occult causes and wonders . . . but never a proof from experiment, never a demonstration do you find in them. The writers . . . treat the subject esoterically; miracle-mongeringly, abstrusely, reconditely, mystically. Hence such philosophy bears no fruit; for it rests simply on a few Greek or unusual terms—just as our barbers toss off a few

> Latin words in the hearing of the ignorant rabble in token of their learning, and thus win reputation . . . few of the philosophers are investigators, or have any first-hand acquaintance with things.[1]

The "miracle-mongerers," by implication, included the Church. It says much for English tolerance that Gilbert was able to publish his views without fear for his life or liberty, especially as he emphatically supported the view of Copernicus that the earth was not the center of the universe. Things were different closer to Rome, where others who advanced Copernican views were brutally dealt with: Giardino Bruni was burned at the stake and Galileo Galilei was kept under house arrest for life.

Spurning the scholars, Gilbert talked to people who *used* magnets: compass makers, navigators, and ship captains. No doubt he heard all the popular theories many times—that magnets were attracted by the North Star or by a huge arctic mountain that would pull out all the ship's iron nails if one got too close to it, and that garlic interfered with compass readings. But an idea came to him that was consistent with all he had heard about how compasses *actually* behaved: Earth could be a giant magnet. To test the idea, he made what he called a "terella"—a model Earth formed from a naturally magnetic iron ore called lodestone—and moved a compass around it. Interpreting the findings as a though he were a traveler on the surface of his surrogate Earth, he found that a compass needle behaved in every way as it did on the real Earth.

To investigate electrical forces, which were much weaker than magnetic ones, he needed a sensitive detector and so made the world's first electroscope. He called it a "versorium"—a light metal needle balanced on a pinhead, rather like a compass except that the needle was not magnetized. When brought toward an electrified object, the needle would move so as to point toward it. Using the versorium, he produced a great list of materials that became electrified when you rubbed them. But this device, wonderful as it was, didn't distinguish positive from negative electricity, so Gilbert failed to discover that some substances became positively electrified and others negatively so. Nor did he notice that two similarly charged objects *repelled* one another. He also failed to see the symmetry of magnetic attraction and repulsion—to him the mutual repulsion exerted by like poles was just a preliminary shuffle to get the unlike poles together.

One can hardly fault him for these failures. Gilbert had made giant strides in the understanding of electricity and magnetism, breaking away from mediaeval thinking and clearing a path to modern science. Among his near-contemporaries were Francis Bacon and Galileo Galilei, even more powerful advocates of what we now call the scientific method. It took Galileo, some twenty years later, to show the full relationship between observation, hypotheses, mathematical deduction, and confirmatory experimentation, but in electricity and magnetism it was Gilbert who showed the way. By reporting exactly what he had done in his experiments, he made it possible for others to repeat them, verify the results, and, perhaps, extend them. Others were led to study the subject, and in the 1620s, Niccolo Cabeo, an Italian teacher of theology and mathematics, found what had eluded Gilbert. He noticed that iron filings seemed to jump away from a piece of electrified amber as soon as they touched it. Like magnetism, electricity could push as well as pull.

Scientific knowledge was advancing, but much of it was still contained in statements of the "if you do so-and-so, then such-and-such will happen" variety. The notion that everything that happened in the physical world might be governed by universal laws in mathematical form seemed as fanciful as the magic mountain. But everything changed in 1687 when Isaac Newton published his *Principia Mathematica*. He showed that three simple laws were sufficient to describe how any material object moved under the action of forces, and that any two objects attracted one another with a force that was proportional to the product of their masses and inversely proportional to the square of the distance between them. Everything from an apple's fall to a planet's orbit could now be described by precise equations. It is hard to find words adequate to describe Newton's achievement. Perhaps Alexander Pope did it best when he wrote an epitaph for Newton in 1727: "Nature and Nature's Laws lay hid in night: God said, Let Newton be! And all was light."

Science had entered a new era and was set on the path that it still follows today. The aim was to bring everything within universal laws—the fewer and the simpler, the better—and to do this by employing both experiment and mathematics. Newton himself did some spectacularly successful experiments on light, showing that what we perceive as white light was actually a mixture of all the colors in the visible spectrum. He never turned his hand to electricity and magnetism, but, as we'll see, others came to use his law of gravitation as a model for both.

Meanwhile, scientists were slowly improving their acquaintance with the ways of electricity. In the 1730s, the French army officer turned chemist Charles du Fay discovered that glass, when rubbed with silk, acquired a different kind of electricity from that acquired by amber when rubbed with fur. Moreover, while electrified glass attracted electrified amber, two pieces of electrified glass repelled one another, as did two pieces of electrified amber. Du Fay thought that the amber and the glass might have each became imbued with a distinct type of electrical fluid and introduced what became known as the "two-fluid" theory. Meanwhile, an American was working along different lines. Benjamin Franklin was someone who, it seemed, could do anything. Already a brilliantly successful printer, publisher, and journalist, he went on to become a distinguished, if rather raffish, politician and statesman. He was also a great scientist. In 1747 he put forward the idea of electrical charge, which could be positive, as with glass, or negative, as with amber. Franklin's charge came in the form of a *single* hypothetical electrical fluid. By his "one-fluid" theory, a body with the normal amount of fluid would have no charge, but one with a surplus of fluid was positively charged and one with a dearth of fluid was negatively charged. To interpret du Fay's experimental findings, Franklin assumed: (1) that rubbing transferred electrical fluid from the silk to the glass but took it from the amber to give to the fur; and (2) that, by analogy with magnetic poles, unlike charges attracted one another while like ones repelled each other. By the same token, this explained why Cabeo's iron filings jumped away from the electrified amber—when they touched it, they acquired the same (negative) charge and were repelled.

According to a popular story, Franklin flew a kite into a thundercloud to prove that lightning was electrical and so establish a sound theoretical base for installing lightning rods on buildings. The experiment was successful and lightning rods became widely used. Franklin, if indeed it was he, had saved many lives by risking his own—several others were killed trying similar experiments. For the kite experiment something was needed to collect the electricity from the lightning and store it for later examination. Such a device had been invented a few years earlier by Pieter van Musschenbroek, a professor at Leyden in Holland. He tried storing electricity in a water-filled jar and was more successful than he had thought possible. While he turned the handle of a machine that generated electricity by mechanical rubbing, his student and assistant Andreas Cunnaeus picked up the jar to try to draw a spark from it to a gun barrel held in his other hand. They had seen sparks before, but not like this one. There was a great flash, and the shock that passed through Cunnaeus's body almost killed him. Their device, improved by coating the jar, inside and outside, with foil and dispensing with the water, became the Leyden jar, the first capacitor. It quickly became the standard equipment for storing electricity, and many people, including young Faraday, later made their own versions of it to use in home experiments.

Gilbert and others had long ago observed that electric and magnetic forces became weaker as one moved away from an electrified body or a magnetic pole. As Newton's inverse-square law worked so well for gravity, it seemed likely that similar laws applied to electricity and magnetism. John Michell demonstrated in 1750 that this was so for magnetism, and Joseph Priestley did the same for electricity in 1766. But it was the French physicist Charles Augustin Coulomb who carried out a definitive set of experiments in 1785 and gave his name to the law. For the purpose, Coulomb had independently reinvented a wonderfully precise instrument—the torsion balance.[2] John Michell had begun to construct one thirty years earlier but had died before putting it to use.

Coulomb didn't hold with Franklin's one-fluid hypothesis. Since du Fay's time, French scientists had come to believe in the two-fluid theory of electricity—that there was one fluid for the amberlike substances and another for those like glass—and, when

Coulomb endorsed it, this model became ingrained in French scientists' thinking. On the other side of the Channel, Franklin was very popular and British scientists became ardent supporters of his single-fluid version. The debate went on for many years, and both sides had passionate adherents. It all seems rather silly and irrelevant now, like an argument about whether flying pigs have one pair of wings or two, but in the late 1700s, so-called imponderable fluids—hypothetical substances undetectable by the senses—were serious components of scientific thinking; for example, Antoine Laurent Lavoisier, the father of chemistry, believed that heat was a fluid called "caloric."

Magnetism similarly had its fluids, two for both the British and the French. But, whether with one fluid or two, the theories of electricity and magnetism now followed the pattern of Newton's law of gravitation, with the difference that electricity and magnetism both attracted and repelled while gravity only attracted. The equations looked just like Newton's and gave exact values for the forces. It all looked so right. But, buried deep, there was a flaw. Newton had seen it. He had written to his friend Richard Bentley:

> That gravity should be innate, inherent, and essential to matter, so that one body can act on another at a distance, through a vacuum, without the mediation of anything else, by and through which their action and force may be conveyed from one to another, is to me so great an absurdity that I believe no man who has in philosophical matters a competent faculty of thinking, can ever fall into it.[3]

Newton knew that his equations were not the last word on the matter. No force could act instantaneously across a distance. *Something* had to exist in the intervening space to transmit the force, even though he carefully avoided making any hypothesis about what it was. But in France, mathematically inclined physicists pushed thoughts about a transmission medium to the back of the mind while confidently building on Newton's foundation. They did not feel the need to agonize about the ultimate meaning of gravity or other forces and how they were transmitted; it was enough to present the mathematical equations describing the universe, and, *voilà*, it all became comprehensible. They believed, as Newton did, that the whole physical world behaved as though composed of point masses that obeyed precise laws (as did the planets) and so reduced the reality of ponderable matter in the universe down to a series of equations. Their mathematics was elegant, elaborate, and all-encompassing, and the results were beautiful.

Joseph Louis Lagrange produced the definitive work on dynamics, his *Mécanique analytique*, which didn't contain a single diagram; Pierre Simon Laplace wrote his masterly, five-volume *Mécanique céleste*; and Siméon Denis Poisson worked out the exact distribution of charge on the surfaces of two spherical conductors a given distance apart. Newton's warning about the "absurdity" of supposing gravity to act instantaneously at a distance without any kind of medium was not entirely forgotten. Supposing the medium to be fluid, Laplace calculated that gravitational forces would be transmitted at least seven million times faster than light, and this result seems to have reassured others that any difference from instantaneous action, both for gravity and for electrical and magnetic forces, could be safely ignored. With electricity and magnetism, it seemed that Newton's gravitational model had provided a royal road, but it eventually turned out to be a cul-de-sac.

Before 1800, all man-made electricity was static. The discovery of continuous currents came as a complete surprise and was in the best tradition of scientific serendipity. Luigi Galvani, an anatomist in Bologna, used to hang dead frogs' legs on a row of brass hooks to dry. When, around 1780, he touched one of the legs with a piece of iron that happened to be in contact with the brass hook, the leg twitched! The muscle tissue in the frog's leg was producing electricity, or so Galvani thought. His friend Alessandro Volta disagreed and set out to test an alternative theory that the electricity was being generated by chemical action between the different metals in contact with the frogs' legs. The result, a decade or so later, was the voltaic pile, or battery. His first battery was a pile of alternate discs of silver and zinc interleaved with layers of brine-soaked pasteboard. Amazingly, this simple assembly of seemingly inert components produced a continuous electric current when the silver disc at one end was connected through a metal circuit to the zinc disc at the other. The more discs in the pile, the greater was the electric effect, and it turned out that there was no need for expensive silver. Any two metals would do; copper and zinc worked very well.

Volta had never intended to invent the battery, but it soon took on a life of its own. Experimenters discovered that interesting things happened when you connected a wire to each of the terminals of a battery and dipped the two wire ends in a chemical solution. Up to now, the only way chemists could investigate the constituents of matter was to mix substances together and watch what happened. Now a new technique called electrochemistry opened up the field. You could make a solution of the substance under investigation, dip the two wires from a battery into it, and let the electrical force do the work; sometimes it would separate the constituent parts of the substance, each part being drawn toward one of the wire ends. As we have seen, news of this eventually reached the young enthusiasts at the City Philosophical Society, and Faraday, using his seven-half-penny battery, joyfully watched as bubbles of gas appeared and his solution of magnesium sulfate turned murky as it decomposed.

At the other end of the size scale, Humphry Davy was using his huge battery, formed of 2,000 voltaic cells, at the Royal Institution to break new ground in chemistry. By bold and sometimes-dangerous experiments, he isolated a string of new elements: barium, calcium, sodium, potassium, magnesium, and boron (the last-named together with the Frenchmen Joseph Louis Gay-Lussac and Louis-Jacques Thénard). In the course of this work, Davy became more and more convinced that *all* chemical reactions were the result of electrical action. His views certainly influenced those of his protégé, though Faraday took nobody's word, even Davy's, for granted and always had to work out everything for himself.

Electricity was shaking up chemistry, but a new wind was coming in from a surprising direction. Immanuel Kant had published his *Critique of Pure Reason* in 1781, and it became a preamble to a German school of scientific thought called *Naturphilosophie*, advanced principally by Friedrich von Schelling. Hans Christian Oersted, professor of physics at the University of Copenhagen, was an adherent of both science and philosophy. The son of an apothecary, Oersted was a close friend of his near-namesake Hans Christian Andersen and brother to Anders Oersted, who became prime minister of Denmark. He was a man of many parts—among other things, he wrote poetry—but he was drawn early on to science, especially chemistry, and to German philosophy.

In his 1786 work, *Metaphysical Foundations of Natural Science*, Kant put forward a dynamical theory of matter—matter as made up of fundamental forces of repulsion and attraction—and thus opened up the possibility of a unified treatment of all forces, including electricity and magnetism. In this theory, all reality was reduced to the two opposing forces of attraction and repulsion. The two forces were diffused throughout all of space and propagated through a medium: by Kant's theory, a medium of some kind was needed to transmit all physical forces—light, gravity, electricity, and magnetism.

Schelling believed, like Kant, that all phenomena could be reduced to the effects of attractive and repulsive forces, but he went further by suggesting that the fundamental forces became manifest in different forms in different circumstances. Electricity, for example, was the manifestation of attraction and repulsion under given physical conditions. According to *Naturphilosophie*, all space was filled by a web of forces that manifested themselves in various forms according to the conditions that existed locally. And along with the unity of all forces went the idea that each form of the force—light, heat, electricity, magnetism, gravitation—could be converted into any of the others under the proper experimental conditions. This was the first step toward the realization of a *field* theory—one in which the energy associated with physical phenomena lies in a continuous medium surrounding bodies rather than in the bodies themselves.

Coulomb, a good Newtonian, argued that material bodies acted on one another at a distance, along a straight line between them, with the intervening space playing no part at all. Unlike Kant and Schelling, Coulomb believed that each type of force was distinct: for example, electrical forces required a different type of fluid from magnetic forces, so it was impossible that one could be converted to the other.

From our distant and privileged viewpoint, it is evident that Kant and Schelling were right, at least in broad terms, and that Coulomb was wrong. But things looked different in the early 1800s. Coulomb's equations were clear, elegant, and gave exact answers, while the ideas of Kant and Schelling were speculative and vague, even metaphysical. After being taken aback by Oersted's discovery in 1820 of the connection between electricity and magnetism, André Marie Ampère, the French mathematical physicist whom Davy and Faraday had met in Paris, explained in a letter to a friend why none of his countrymen had thought of placing a compass near a current-carrying wire, as Oersted had done:

> You certainly have a right to ask why it is inconceivable that no one tried the action of the voltaic pile on a magnet for twenty years. However, I believe that the cause of this is easily discovered: it simply existed in Coulomb's hypothesis on the nature of magnetic action; everyone believed this hypothesis as though it were a fact; it simply discarded any possibility of the action between electricity and so-called magnetic wires. . . . Everyone resists changing ideas to which he is accustomed.[4]

Oersted approached electricity and magnetism from quite a different angle. He had been strongly influenced by *Naturphilosophie*, along with its tenet of the unity of all nature's forces. In principle, he took the Kantian view that all matter was made up of two forces, one that attracted and one that repelled, but he interpreted these as combustion and combustibility. When latent, these forces constituted the chemical properties of bodies, and when conditions allowed them to act freely they produced electricity. An idea formed for exploring how electricity and magnetism might be connected. He would use a battery to send an electric current through a thin wire filament that would grow hot and glow—he thought that magnetic effects might eradiate from the wire along with heat and light. Near the wire would be a compass; perhaps its needle would be deflected. He built the apparatus and tentatively tried it during a lecture. When he connected the battery, the needle twitched, but very feebly, and the experiment made little impression on the audience. Nor can it have done on Oersted because three months passed before he tried it again.

Once more, just a feeble twitch. Only when he used a thick wire in place of the filament, thereby greatly increasing the current, did the needle take up a new fixed position. It set itself at right angles to the wire! This time there was no heat and no glow: the connection was directly between electricity and magnetism. The electric current set up a "conflict" that acted *in a circle* around the wire and gave rise to a force when it encountered magnetic materials.

This effect was like nothing seen before, and it sent shock waves through the scientific community. Oersted's own explanation of it, though understandably somewhat tentative and vague, ran completely counter to the accepted Newtonian doctrine of push-pull forces acting at a distance in a straight line, and so added to the shock. He wrote:

> We may now make a few observations towards explaining these phenomena. The electrical conflict acts only on the magnetic particles of matter . . . their magnetic particles resist the passage of this conflict. Hence they can be moved by the impetus of the contending powers.
>
> It is sufficiently evident from the preceding facts that the electrical conflict is not confined to the conductor, but dispersed pretty widely in the circumjacent space . . . we may likewise collect that this conflict performs circles; for without this condition it seems impossible that the one part of the uniting wire, when placed below the magnetic pole, should drive it towards the east and when placed above it towards the west; for it is the nature of a circle that the motions in opposite parts should have an opposite direction. Besides, a motion in circles, joined with progressive motion according to the length of the conductor, ought to form a conchoidal or spiral line, but this, unless I am mistaken, contributes nothing to explain the phenomena observed.[5]

Oersted had jolted physical science on to a new track. Everyone now knew that electricity and magnetism were inextricably linked. But the exact nature of the linkage was to prove elusive. Finding it would take boldness, tenacity, and genius.

CHAPTER FOUR

A CIRCULAR FORCE

1820–1831

Davy and Faraday lost no time in repeating Oersted's experiment. They watched the magnetic needle set itself at right angles to the current-carrying wire and, by setting the wire vertical and moving the needle around it, found that the force did, indeed, act in a circle. Seeing is believing, but there must have been a moment when they doubted their senses. Nature's three known primary forces—gravity, electricity, and magnetism—either pulled directly toward the source or pushed directly away from it, in accordance with Newtonian principles. This new force acted *sideways*.

At this time, Faraday was immersed in laborious experiments on alloys of steel—bringing in much-needed income for the Royal Institution—and had another serious distraction that we'll come to shortly, so Davy turned to his friend William Hyde Wollaston for help. Wollaston was a distinguished scientist who, among many achievements, had demonstrated that electricity produced by friction, for example by rubbing glass with silk, is the same as that produced by a battery. Together, they quickly dismissed Davy's original guess that the current-carrying wire itself became magnetized, and Wollaston began to carry out some promising experiments.

Meanwhile, someone else was blazing a trail in Paris. André Marie Ampère latched onto Oersted's discovery with astonishing speed. In an unmatched display of scientific virtuosity, he produced a theory of electromagnetism in only a few months that went on to win almost universal acceptance. And it wasn't just a paper theory; he backed it up with exquisitely elegant experiments. Ampère had overcome personal tragedy to become a popular professor at the École

Polytechnique—his father had been a victim of the terror inflicted by the Jacobins in the wake of the Revolution and his wife had died after only four years of marriage. A staunch member of the French Newtonian school, he sought a way to explain Oersted's discovery in terms of straight-line forces and action at a distance, and, with wonderful ingenuity, he found one.

His inspiration was to test whether two parallel current-carrying wires exerted a force on one another. To his, delight they did—the wires were attracted to one another when the currents ran in the same direction and repelled when the currents ran in opposite directions. The momentous idea came to him that electric currents might be the source of all magnetism. How, then, did permanent iron magnets work? His first idea was that cylindrical currents whirled around an axis that ran between the north and south poles of the magnet, but, if so, where did the currents come from and why had nobody noticed them? In the course of trying to detect these currents, Ampère's friend Augustin Fresnel made the astonishingly prescient conjecture that magnetism in iron was produced by currents circulating around each minute particle of metal, each loop of current acting as a tiny magnet. In a permanent magnet, these circulating currents became aligned with one another and so acted cumulatively to produce a strong magnetic effect.

Ampère conjectured that the total effect of all these little internal currents in a cylindrical-shaped permanent magnet would be the same as that of a single current circulating only on the surface of the cylinder. The idea was easily tested. He wound a wire in the form of a helix and passed a current through it. His helical coil—the world's first solenoid—was, in effect, carrying a cylindrical current and it behaved exactly like a permanent magnet of the same size and strength. When carefully suspended, it aligned itself with Earth's magnetic field, just like a compass needle.

Like almost all scientists of the time, Ampère, was steeped in the Newtonian tradition. As he saw it, the pressing question was how to explain his newly discovered force between current-carrying wires in Newtonian terms. His mathematical skill now came to the fore.

An electrical circuit could be treated mathematically as a string of infinitesimal current elements, each with its own strength and direction. Any two of these elements would exert a mutual force of attraction or repulsion that would depend on their strength and direction and could be assumed to act along the straight line joining them. Moreover, true to Newton's precepts, the force would be inversely proportional to the square of their distance apart. (Ampère had shown by experiment that such a law applied to two current-carrying circuits in parallel planes, and it was reasonable to assume that all elements of the wires behaved in the same way.) By this means, Ampère produced a formula for the mutual force between any two current elements, no matter what their strength and direction or their spatial separation The total force between two complete current-carrying circuits could then, in principle, be worked out by summing the forces between every pair of elements mathematically.

Ampère's work was a tour de force. Faraday would soon form his own ideas on the subject, but in the winter of 1820 and 1821, he had other things on his mind. One of his friends at the City Philosophical Society was a fellow Sandemanian Edward Barnard. When Faraday met Edward's nineteen-year-old sister Sarah, he was smitten. A few years earlier, he would have claimed that his single-minded pursuit of scientific truth left little room for the fair sex and that love was something to be avoided. In fact, he had jotted a poem in his commonplace book that began:

> What is the pest and plague of human life?
> And what is the curse that often brings a wife?
> 'tis love.[1]

How empty and foolish those words seemed now. He had found his life's companion, and he opened his heart to her in a letter:

> You know my former prejudices and my present thoughts—you know my weaknesses, my vanity, my whole mind; you have converted me from one erroneous way, let me hope you will attempt to correct what others are wrong. . . . In whatever way I can best minister to your happiness, either by assiduity or by absence, it shall be

> done. Do not injure me by withdrawing your friendship, or punish me for aiming to be more than a friend by making me less; and if you cannot grant me more, leave me what I possess, but hear me.[2]

When Sarah showed the letter to her father, he straightaway packed her off to stay with her sister on the Kent coast. Disconsolate but undeterred, Faraday left his work and followed. For a while he made no headway in his quest, but then came a wonderful day when he and Sarah took a tour of the cliffs of Dover. His journal entry brims with joy.

> The cliffs rose like mountains. . . . At the foot of these cliffs was the brilliant sparkling ocean, stirred with life by a fresh and refreshing wind, and illuminated by a sun which made the waters themselves seem inflamed. . . . I can never forget this day. Though I had ventured to plan it, I had little hope of succeeding. But, when the day came, from the first waking moment in it to the last it was full of interest to me; every circumstance bore so strongly on my hopes and fears that I seemed to live with thrice the energy I had ever done before.[3]

Sarah had accepted his proposal. They were married in June 1821 and remained devoted to one another all their lives. Though Sarah knew nothing of science, she understood very well how important her husband's work was to him and to the wider world. She made sure that he ate nourishing food and did her best to see that the long hours in the laboratory were balanced by adequate relaxation. She was, as Faraday put it, "a pillow to his mind."[4] They had no children of their own, but their apartment at the Royal Institution was a vibrant, happy home, often the scene of family celebrations at birthdays or after weddings. It was always a treat for young relations to visit Uncle Michael and Aunt Sarah: There were lively games and a never-ending supply of soda, ginger wine, and lozenges that Faraday made in the laboratory downstairs. Even there, he would let the children sit and watch him work. Sometimes he would put on a little show for them—tossing some potassium into water so it would fizz and dart around amid lilac-colored flames, or sealing up some mercury in a glass tube so they could feel its weight and watch it roll around.

Back at work after the marriage, Faraday had a request from one of his friends at the City Philosophical Society. Richard Phillips had been appointed editor of the journal *Annals of Philosophy*, and he asked Faraday to write a historical account of electromagnetism. Faraday had, so far, only dabbled in the subject, but this was an important commission and he brought all his faculties to bear. He read everything he could find, repeated the experiments, and did his best to follow the reasoning of Oersted and Ampère.

As we now know, Oersted's view, derived from Kant, that all space was crisscrossed by forces of one kind or another, turned out to be somewhere near the truth. It was also close to the view that Faraday eventually came to. In all likelihood, he had also been indirectly influenced by Kant, Schelling, and *Naturphilosophie*. Davy's poetical and philosophical friend Samuel Taylor Coleridge had become an evangelist for *Naturphilosophie* after visiting Germany in 1798, and some of his enthusiasm had rubbed off on Davy, in particular the notion of the unity of all nature's forces. Faraday, in his formative years, had thus been exposed to the ideas of the German school. But, try as he might, he couldn't make heads or tails of Oersted's vague theory of "conflicts," nor of his proposition that an electric current was a wave of chemical disruption and reconstitution. Ampère's work was much more to the point. It was precise and elegant, and the mathematically based theory was backed by experiment. But Faraday's self-education was deficient in one significant respect: he had learned no mathematics. For him, Ampère's equations might as well have been written in Egyptian hieroglyphics.

We shall never know what Faraday would have achieved had he mastered mathematics, but, paradoxically, his ignorance may have been an advantage. It led him to derive his theories entirely from experimental observation rather than to deduce them from mathematical models. Over time, this approach gave him a deep-seated intuition into electromagnetic phenomena. It enabled him to ask questions that had not occurred to others, to devise experiments that no one else had thought of, and to see possibilities that others had missed. He thought boldly but would never commit himself to an opinion until it had withstood the most rigorous experimental testing. As he explained in a letter to Ampère:

> I am unfortunate in a want to mathematical knowledge and the power of entering with facility any abstract reasoning. I am obliged to feel my way by facts placed closely together.[5]

For Ampère, on the other hand, mathematics was the language of nature. As we've seen, his views on electricity and magnetism were derived mostly by mathematical analogy with the theory of gravitation. He did carry out some fine experiments, but these were largely to confirm theories he had already developed by abstract reasoning. The stark difference between the two great scientists stemmed from their backgrounds. Ampère was a product, and now a leading member, of the well-established French Newtonian school of mathematical physicists, whereas Faraday, although nurtured under Davy's patronage, was very much his own man—the outsider who eventually came to take center stage. Despite their differences, the two men were drawn together by a mutual passion for science, and they enjoyed many years of friendly correspondence. Faraday believed that differences of opinion served to ferret out the truth.

Much as he admired Ampère's work, Faraday began to develop his own views on the nature of the force between a current-carrying wire and the magnetic needle it deflected. Ampère's mathematics (which he had no reason to doubt) showed that the motion of the magnetic needle was the result of repulsions and attractions between it and the wire. But, to Faraday, this seemed wrong, or, at least, the wrong way around. What happened, he felt, was that the wire induced *a circular force* in the space around itself, and that everything else followed from this.

The next step beautifully illustrates Faraday's genius. Taking Sarah's fourteen-year-old brother George with him down to the laboratory, he stuck an iron bar magnet into hot wax in the bottom of a basin and, when the wax had hardened, filled the basin with mercury until only the top of the magnet was exposed. He dangled a short length of wire from an insulated stand so that its bottom end dipped in the mercury, and then he connected one terminal of a battery to the top end of the wire and the other to the mercury. The wire and the mercury now formed part of a circuit that would remain unbroken even if the bottom end of the wire moved. And move it did—in rapid circles around the magnet!

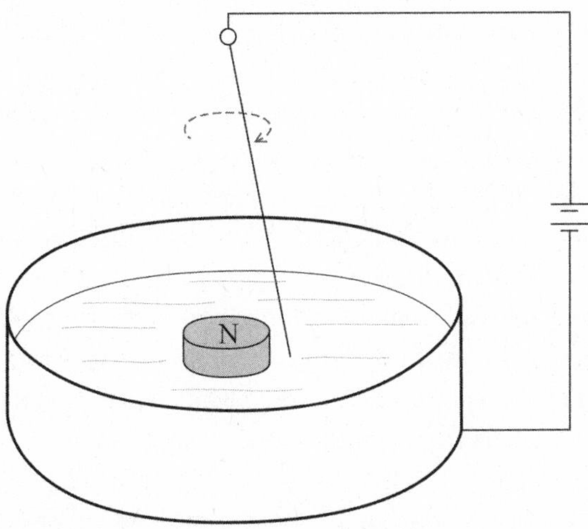

Fig. 4.1. Faraday's first electric motor apparatus. (Used with permission from Lee Bartrop.)

Not done yet, he modified the apparatus slightly, freeing the magnet and letting it float in the mercury, but with one end tethered to a fixed point in the base of the basin. About a quarter of the magnet was now exposed above the surface of the mercury. He replaced the dangling wire with a fixed one that dipped into the mercury at the center of its surface, and then he reconnected the battery. This time, the magnet revolved around the wire! Faraday had become a discoverer: he had made the world's first electric motor. He and George danced around the table and went off to the circus to celebrate. George later recalled the moment: "I shall never forget the enthusiasm expressed in his face and the sparkling in his eyes."[6] We can imagine the joy with which Faraday wrote the simple words in his journal: "Very satisfactory, but make a more sensible apparatus."[7]

He might have added: Get it published—tell the world about it. (Faraday's constant motto was Work, Finish, Publish.) With that thought in mind, he dashed off a paper called "On Some New Electromagnetic Motions and the Theory of Electromagnetism" in time to catch the next edition of the *Quarterly Journal of Science*. Within a month, his discovery was in print; but within another week, his elation vanished. In his haste, he had neglected to pay the customary compliments to his mentor and senior colleague Davy. Worse than that, he had failed to mention the work of Davy's close friend, Wollaston, who had been trying for a year to produce rotary motion with currents and magnets. Though Faraday had not worked with Davy on electromagnetism since Wollaston came on the scene, he had overheard the two in conversation and had a rough idea of what Wollaston was doing. In fact, Wollaston was on another track—trying in vain to get a wire coil to spin around its own axis in response to a magnet—but the difference between that kind of rotation and Faraday's was too subtle for the casual observer to recognize. Faraday was accused of plagiarism, not by Wollaston directly, but by others, including Davy.

To be suspected of such a dishonorable act was a sickening blow, made worse by knowing that Davy, the man he most admired in the world, was his leading accuser. We can only guess at Davy's motives. A complex character, he was both generous and vain. Perhaps the best, simple explanation is that generosity prevailed while Faraday was a protégé whose achievements boosted his own public standing, but vanity took over when the protégé appeared in the character of a rival. And Faraday's ill-mannered oversight in not sharing some of the credit with his guide and mentor was an affront to Davy's dignity.

Desperate to clear himself of the slur, Faraday wrote to Wollaston to apologize. He received a somewhat dismissive reply:

> You seem to me to labor under some misapprehension of the strength of my feelings on the subject to which you allude. As to the opinions which others have of your conduct, that is your concern and not mine; and if you acquit yourself fully of making any incorrect use of the suggestions of others, it seems to me that you have no occasion to concern yourself much about the matter.

Faraday battled alone to rebut the plagiarism charge and, for the most part, succeeded. Even so, he was still out of favor with the old guard of the British scientific establishment who set great store by protocol and expected deference from their juniors.

But the wider world cared little for such things, and his discovery took on a life of its own. Within a few months, the Royal Institution had a large rotator in its lecture hall for all to see and had sent pocket-sized versions to scientists around Europe. Electromagnetism was now a hot topic. Many people wanted to find out more about it and turned gratefully to a series of articles in the *Annals of Philosophy* that set out everything for them. The author had modestly called himself "M." Rumors spread, and they pointed overwhelmingly to Faraday. This was, indeed, the historical review he had written at Richard Phillips's request, and he was obliged to acknowledge authorship. He was famous.

The next step in his career was to become a fellow of the Royal Society, and a band of current fellows, including Wollaston, put him up for election. Things seemed to be going well, but the prelude to the election turned out to be one of the most unpleasant episodes in his life. The accusations of ungentlemanly conduct had not gone away, and there was opposition from a small group headed by the president, Sir Humphry Davy. Faraday was in a corner. To overcome the injustice, he was forced to put aside all the rules of his Sandemanian upbringing and press for his own advancement by appealing to those who stood in his way. This was a hateful task, but he carried it out, and in January 1824, he was elected F. R. S., with one dissenting vote. It is a measure of Faraday's character that he never harbored rancor about this incident, although he did confess to a friend that his relationship with Davy was never the same afterward.

The following year, Davy nevertheless instigated Faraday's promotion to director of the Royal Institution—perhaps it was a case of *force majeure*. Faraday's immediate task was to rescue the Institution from a precarious financial state, and he did it with panache rivaling that of Davy twenty years before. On Davy's home turf, too—the lecture hall. He started by inviting the Institution members to talks in the laboratory on Friday evenings, but these soon became so popular that he moved them to the grand lecture theater upstairs. So began the tradition of Friday Evening Discourses at the Royal Institution that continues today. Even the format is unchanged, theatrical in its simplicity—at precisely the appointed time, the speaker enters unannounced, speaks for exactly one hour, takes his bow, and leaves the stage. The early talks went so well that Faraday decided to put on a special set of lectures for children at Christmas. These, too, have run ever since; the television audience for them now is huge. Faraday's immediate purpose was served, too: the Institution's membership, and its income, swelled as people were drawn to the lectures. Though by no means prosperous, the Institution now at least had its head above water.

The success of the talks didn't happen by chance. As we've seen, Faraday had early on begun to form his own ideas on the art of the scientific lecture, and had, by now, built up an unrivaled body of expertise: had he written a book on the topic, it would have become the standard work. His various notes contain guidance on everything from the layout of the seating to the ventilation and lighting in the hall, and copious advice to the lecturer, beginning: "A flame should be lighted at the commencement and kept alive with unremitting splendour to the end."[8] Faraday gave many of the lectures himself and drew devoted audiences just as Davy had done, though with quite a different style. In place of Davy's flamboyance and flashes of brilliant improvisation there was simple charm, as he conveyed the wonder of science with consummate phrasing and timing. He soon became the foremost public lecturer on science in England in a career lasting from 1823, when he was called in unexpectedly to substitute for William Brande, to his final appearance in 1862.

The nature of Faraday's genius is hard to pin down, but his mastery of the lecture offers some clues: his knowledge of the subject came not from what he had read or been told but from personal observation; he looked and listened with rare intensity and was able to capture subtle effects and nuances that passed other people by. By the same token, he took pains to identify all the factors that contributed to an audience's enjoyment and evaluated their effects, both singly and in combination. Perhaps the most revealing aspect is that he never demonstrated an experiment on stage, no matter how spectacular, unless he could also present the audience with the theory behind it. His scientific genius lay not simply in producing experimental results that had eluded everyone else but in explaining them, too.

Faraday's elevation to the scientific establishment led to new demands on his time. While coping with the day-to-day business of the Royal Institution, he had to fend off requests to take on other administrative work, for example secretaryship of the new Athenaeum Club. But there was one request he couldn't refuse. In the early 1700s, Britain had led the world in the manufacture of high-quality glass for optical instruments, but in 1746, the government decided to raise money by levying a heavy tax on all glass. The goose that had laid the golden eggs slowly died from suffocation—by the 1820s, the French and the Germans were making superb lenses, but British manufacturers had forgotten how to do it. The government couldn't bring itself to drop the tax, but in 1825, the Royal Society set up the Committee for the Improvement of Glass for Optical Purposes to try to rescue the situation. Faraday was invited to join—in effect to run the project—and it was his patriotic duty to accept.

In a backbreaking series of experiments—first at the nearby Falcon Glass Works, then in his own laboratory using a freshly installed glass furnace—Faraday examined all possible causes of imperfections in the glass, eliminating them one by one, at the same testing new methods and ingredients. It was tiresome work that dragged on and on; after each failure, the process of finding the cause and putting it right took weeks. All this was fitted in with Faraday's other duties, and whatever he could squeeze in of his own research. After three years, he succumbed to what he described as "nervous headaches and weakness"[9] and Sarah took him off to the country for two months to recover. This sequence of hard work, mental exhaustion, and enforced relaxation was one that was to repeat itself several times during his career.

Possibly nobody but Faraday could have wrought any success from this thankless task, but in 1830, he delivered a modest-sized sample of acceptable glass, in which he had used a silicated borate of lead in place of the more traditional lead oxide. It worked well when used in a telescope lens, and what the committee wanted now was more of it, to make bigger lenses. Faraday felt himself being sucked into a morass. If he didn't wrench himself free now, he would spend much of the rest of his life working for committees. In 1831, he parceled up six volumes of experimental notes, sent them to the Royal Society, and resigned from the optical glass committee.

Thirty-nine-year-old Faraday had done his duty. After years of frustration through having too little time to devote to his own research, he took matters into his own hands. He had already turned down the offer of a professorship at the new London University but accepted a part-time post at the Royal Military Academy at Woolwich—he enjoyed teaching, and the £200 annual salary was a useful boost to his modest pay from the Royal Institution. But now there would be no more work for committees, no more hack analysis work for commercial companies even though that could have made him a rich man. And no striving for high office. The Royal Institution was both his home and his professional stage; he would stay there and follow his scientific muse. There was one good legacy from the wearisome glass project. Faraday had acquired an assistant—Sergeant Anderson, newly retired from the Royal Artillery—who remained in post until he died in 1866. Faraday's successor at the Royal Institution, John Tyndall, who had a great admiration for Anderson, summed up his merits in two words: "blind obedience."[10] Faraday's friend Ben Abbott was fond of telling a story that one night Faraday forgot to tell Anderson he could go home and came in the following morning to find him still stoking the furnace.

Ten years had passed since Faraday had produced rotary motion with an electric current and a magnet, but the mystery of it had been a constant backdrop to his thoughts. He now had no close colleagues in England, but he enjoyed a lively and friendly correspondence with Ampère. They liked each other and had huge mutual respect even though they disagreed fundamentally about nature's mechanism for electromagnetism. In fact, Faraday was beginning to realize how sharply his own ideas diverged from the mainstream. Ampère's highly mathematical theory of action at a distance, set in the French Newtonian tradition, seemed to give leading scientists all they needed and was almost universally accepted. Faraday acknowledged this in his own writings, where any contradictory suggestions were very cautiously expressed. The reason he had uncharacteristically used a pseudonym in his review of electromagnetism for the *Annals of Philosophy* was probably that he didn't want to be characterized as a bumptious upstart. His letters to Ampère, though, were free and frank; for example:

> I am naturally sceptical in the matter of theories and therefore you must not be angry with me for not admitting the one you have advanced immediately. Its ingenuity and applications are astonishing and exact but I cannot comprehend how the currents are produced and particularly if they be supposed to exist round each particle and I wait for further proofs of their existence before I finally admit them.[11]

Ampère had become a trusted confidant, as is evident from another letter, in which Faraday contrasts their working lives:

> Every letter you write me states how busily you are engaged and I cannot wish it otherwise knowing how well your time is spent. Much of mine is unfortunately occupied in very common place employment and I may offer this as an excuse (for want of a better) for the little I do in original research.[12]

There were, indeed, many distractions from the work he really wanted to do. Once, when immersed in investigations into electromagnetism, he had to put everything aside to test thirty-two samples of the Royal Navy's oatmeal to determine whether they were contaminated. Nevertheless, in whatever time he could spare from other duties, Faraday had got on with exploring electricity and magnetism by experiment. It was probably through testing out Ampère's ideas that his own began to develop. He started by bending a current-carrying wire into a circular loop and found, as Ampère had done, that the loop of current behaved exactly like a magnet—its south pole was on the side from which the current appeared to flow clockwise, and its north pole was on the other. With a single loop, the force was feeble; but by winding the wire into a helix with many turns, he made a powerful magnet. By Ampère's theory, the magnetic force was simply what you got when you added all the straight-line forces between pairs of current elements mathematically. Faraday saw things differently—to him, the magnetic force that curved around any current-carrying wire was not an indirect, mathematically derived effect of straight-line forces, it was something primal, a circular force in its own right. The idea of a circular force was quite beyond the generally accepted doctrine of Newtonian forces, and Faraday's lack of a traditional scientific education probably made it easier for him to accept it. His thinking ran on in a way even further removed from the Newtonian model—he reasoned that by winding the wire into a helix, he had squashed parts of the circular force into a tube that ran through the helix and allowed the other parts of the force to spread out into space.

Whatever theories about circular magnetic forces acting through space had begun to form in his mind, he had, so far, kept such speculations to himself—to publish them would be to court ridicule. More investigation was needed, and he devised a typically ingenious experiment, one that seems simple only *after* you have thought of it. He wound a helical wire coil around a glass tube and connected the coil to a battery, so forming a magnet. With his coil in a fixed horizontal position, he submerged it halfway in water. Then he took a magnetic needle as long as the helix, pushed it through a thin cork, floated the needle-bearing cork on the water, and turned on the current. He now had two magnets: the half-submerged current-carrying coil, which was in a fixed position, and the floating magnetic needle, which could move freely. Unlike poles attract, so the north pole of the needle was attracted to the south pole of the coil magnet and moved toward it. By the accepted theory of magnetic poles, it should have stopped when it reached the south pole of the coil magnet—the mutually attracting north and south poles would then be together. But, instead, it kept on moving right through the glass tube and came to rest with its own north pole beside with the *north* pole of the coil magnet. And at the other end, the two south poles were together.

Simple yet profound. By Faraday's interpretation, the experiment had shown that magnetism was not simply a matter of poles attracting or repelling one another. Indeed, it had shown that magnetic forces did not begin and end at the poles but ran in continuous loops all the way through the magnet. Despite the strict professional skepticism that Faraday always applied to his own ideas—he spent much time devising experiments that would show errors in his thinking—he was gradually coming to believe that magnetic forces actually had a *presence in space*. This was another idea that was completely foreign to adherents of the French Newtonian school; they believed that forces were produced by the instantaneous action of one material body on another at a distance, and they didn't concern themselves about what was happening in the intervening space.

There was a new phenomenon that baffled everyone. In Paris in 1825, François Arago had noticed that compass needles were sometimes deflected when a piece of copper moved nearby. To investigate, he suspended a magnetic needle over a rapidly spinning copper disc. Amazingly, the needle rotated, too. Copper was a nonmagnetic material, so what was going on? The answer turned out to depend on a completely new effect that was to be one of Faraday's greatest discoveries. Along with other scientists across Europe, Faraday was intrigued by Arago's result, and, during the long hours working on glass, thoughts on electromagnetism had been gestating at the back of his mind. Sometimes he would pull from his pocket a little, wire-wrapped iron cylinder and gaze at it. One question dominated all others. If electricity could produce magnetism, shouldn't magnetism be able to produce electricity? In his meager free time, he had tried various arrangements of magnets and circuits, so far to no avail. Now he had shaken off the burdensome "common place employment" and could channel all his energy, skill, and experience into solving this mystery.

Humphry Davy had died in 1829 after a debilitating illness. For all Davy's faults, Faraday revered him. They had spent much time together and formed a strong bond. More than just a mentor, Davy had been a friend and an inspiration to Faraday. Faraday felt privileged to have been close to such a great man. When Davy's biographer, John Ayrton Paris, asked to borrow the first letter Davy had sent him, Faraday obliged, with the note:

> I send you the original, requesting you to take great care of it, for you may imagine how much I value it.[13]

The pupil was about to embark on a journey of discovery that surpassed even that of the master. In August 1831, Faraday wrote in his laboratory journal the first words for a new project that was to become his finest work. His *Experimental Researches in Electricity*, a monumental opus written entirely in words without a single formula, had begun.

CHAPTER FIVE

INDUCTION

1831–1840

The Royal Institution's most treasured possession is something that would not look out of place on a municipal garbage dump—a ragged object resembling a cloth-wrapped deck tennis quoit that has become entangled in wire. It is, nevertheless, priceless—one of the most important pieces of scientific apparatus ever made.

By 1831, everyone knew how to make magnetism from electricity. A current-carrying coil would, if carefully suspended, align itself north–south, and an ordinary piece of iron could be transformed into a permanent magnet by placing it briefly inside the coil. Surely it should be possible to do the opposite: make electricity from magnetism. Many had tried, but all had failed. Faraday's own attempts with various configurations of circuits and magnets had fared no better than anyone else's but his thoughts ran on, aided by the little, coil-wrapped iron cylinder that he carried with him and contemplated from time to time.

Fresh inspiration came from another branch of physics. Faraday's love of music had led to a particular friendship with fellow scientist Charles Wheatstone, whose family business made and sold musical instruments. Wheatstone had invented the kaleidophone, a device that reflected light from the tip of a vibrating metal rod mounted on a wooden board. The moving reflections formed complex patterns when projected onto a screen, which delighted Faraday. Wheatstone introduced him to the work of the German physicist and musician Ernst Chladni, who had demonstrated a similar effect with vibrating plates: he found that if you spread sand thinly on a glass plate and stroke the edge of the plate with a violin bow, vibrations in the plate would form the sand into beautiful patterns—graphic representations of the standing waves in the glass.

Fig. 5.1. Chladni figures—patterns formed in sand spread thinly over vibrating glass plates. (Used with permission from Lee Bartrop.)

Moreover, the patterns could be *induced* on the sand-strewn plate by stroking another plate a short distance away—vibrations in the first plate produced sound waves in the air that then caused the second plate to vibrate. As always, Faraday tried it for himself and explored every avenue by varying the conditions of the experiment. He was rewarded with an even more vivid demonstration of acoustic induction—when he poured a mixture of egg white, oil, and water on the second plate instead of sand, the vibrations showed up as very fine striations, a kind of crimping of the liquid mixture.

Steeped in musings about acoustic vibrations and waves, Faraday began to think that electricity and magnetism might be transmitted by waves resembling those of sound, or of light. To test the idea, he decided in the summer of 1831 to link two electric circuits magnetically and see whether sending a current through the first circuit would cause some kind of vibration or wave that would act through an iron magnet to induce a current in the second. He needed the strongest possible magnetic link between the circuits and so commissioned a wrought-iron ring, such as his father might have made, six inches in diameter. He also needed coiled circuits with many turns—each extra turn would increase the magnetic effect of a current. So when the ring arrived, he wound two wire coils around it on opposite sides of the ring, each with as many turns as he could fit in on one layer, followed by other layers on top. To insulate each turn of the coils from neighboring turns, he interposed lengths of string; and to insulate each layer of coil from the next, and from the iron ring, he used sheets of cloth. It was not elegant, but handsome is as handsome does.

He connected one coil (A) to a battery and a switch. This formed the primary, or sender, circuit; its role was to magnetize the iron ring. He hoped that some kind of vibration or wave would act through the iron to induce a current in the other coil (B), which was connected by two long wires to a galvanometer (G) with a light, delicately balanced magnetic needle. (The galvanometer worked on the same principle as Oersted's original method of detecting a current, but it had been much refined by Ampère and others.) Coil (B), the long wires, and the galvanometer formed the secondary, or receiver, circuit.

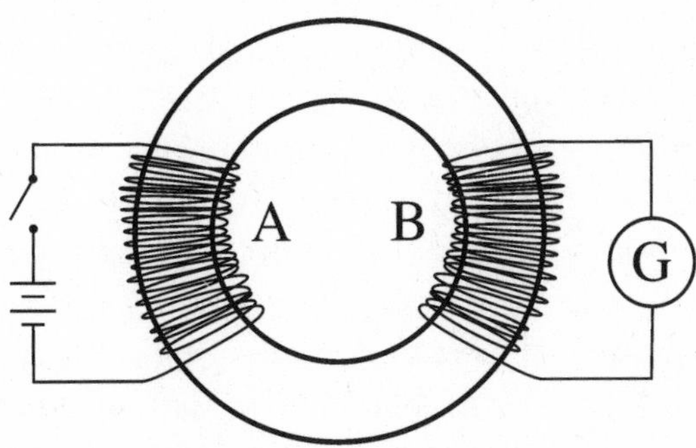

Fig. 5.2. Schematic layout of Faraday's iron ring experiment. (Used with permission from Lee Bartrop.)

If there was even the smallest current, he would see the needle move from its normal north–south position. On August 29, 1831, all was set.

Faraday closed the switch and watched the needle. His heart must have leapt when it moved, but after a brief twitch, it settled back to its resting position. Now a steady current was flowing in the primary circuit but absolutely nothing was happening in the secondary one; the detector needle remained motionless. But when he turned the primary current off, the needle twitched again, this time in the other direction. Strange, indeed, and there was another oddity: the movements of the needle had shown that the first pulse of current in the secondary circuit was in the *opposite direction* to that in the primary, but the second was in the same direction as the primary current.

What could he make of this? It could be one of the greatest scientific discoveries ever, or it could be some chance result with an elusive but mundane explanation. At any rate, Faraday couldn't contain his excitement and wrote to his friend Richard Phillips: "I am busy just now on electromagnetism and think I have got hold of a good thing but can't say; it may be a weed instead of a fish that after all my labor I may pull up."[1]

It was no weed. He repeated the iron-ring experiment many times with minor variations and looked for any possible stray effects that might be causing the needle to twitch. There were none.

No wonder it had taken so long to find a way of generating electricity from magnetism: nothing happened except when the magnetic influence was *changed* by turning the current on or off. A momentous result, but there was more to do. He had generated electricity from magnetism, but so far only from magnetism that he had in turn made from electricity. Was it possible to do the same using only an ordinary permanent magnet and a wire circuit? Various arrangements of magnets with straight wires, helical coils, and spirals yielded no detectable result, but persistence was rewarded when he put a coil-wrapped iron cylinder between the unlike poles of two similar bar magnets and brought the magnets' other ends together to make, in effect, a single V-shaped magnet of twice the strength.

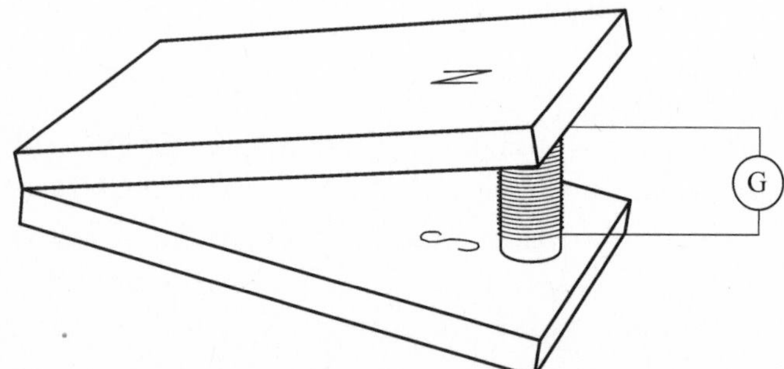

Fig. 5.3. Schematic layout of Faraday's V-magnet experiment. (Used with permission from Lee Bartrop.)

At the instant he brought the magnet ends together, the needle of a galvanometer connected to the coil flicked. And when he pulled the magnet ends apart, the needle flicked the other way. He had generated electricity directly from a magnet and soon found out that there was a much easier way to do it: When he connected a coil of many turns to a galvanometer, took an ordinary magnet, pushed it into the coil cavity and pulled it out again, the galvanometer needle swung vigorously, first one way then the other.

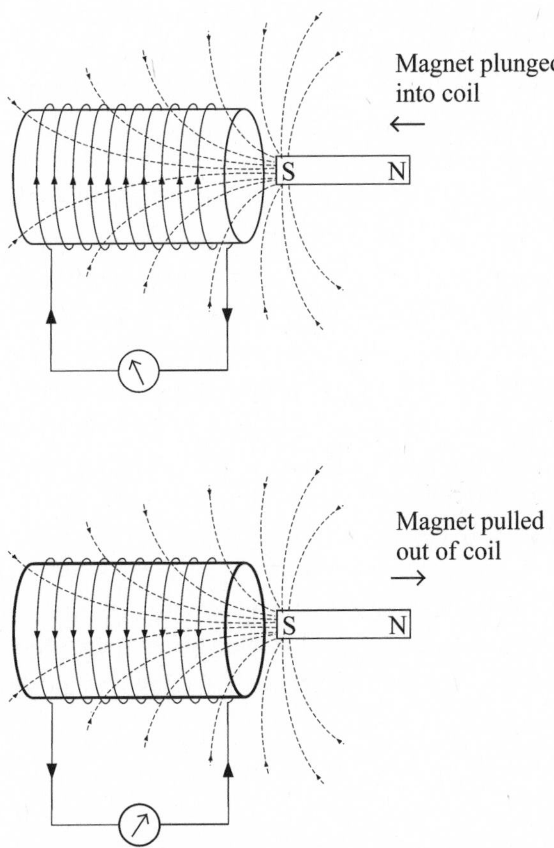

Fig. 5.4. Faraday's magnet-and-coil experiment. (Used with permission from Lee Bartrop.)

He had discovered something that would eventually transform people's lives the world over—electromagnetic induction. To generate electricity in a wire, all you had to do, in principle, was to bring it close to a magnet and set the wire and magnet in relative motion. Faraday had opened up the possibility of generating cheap electricity for any purpose. But little blips of electric current were not enough; what was needed was a continuous current. The blips had been produced by the jerky relative motion of a wire coil and a magnet. Was it possible to devise a smooth motion that would produce a steady current? Faraday's thoughts turned again to the curious experiment conducted by François Arago seven years earlier. Arago had made a magnetic needle spin by rotating a nearby copper disc. Nobody had yet explained why this happened, but Faraday now thought he might have the answer. Perhaps the relative motion of Arago's disc and needle had produced electric currents in the disc—if so, the magnetic effect of these currents would have sent the needle spinning.

Faraday decided to try a variation on Arago's experiment. He mounted a copper disc on an axle and set its edge in a narrow gap between the poles of a powerful magnet. He then made an electrical circuit by placing one sliding contact on the edge of the disc, placing another on the axle, and connecting the two contacts by wires to a galvanometer. He hoped that when the disc rotated, its motion relative to the magnet would produce a steady current across the disc that would register on the galvanometer. He spun the disc and watched the needle. It moved, and this time stayed in its new position; there was a feeble but steady current. And when the disc was spun the other way, the needle also reversed its movement. Ten years after making the world's first electric motor, Faraday had made the world's first dynamo.

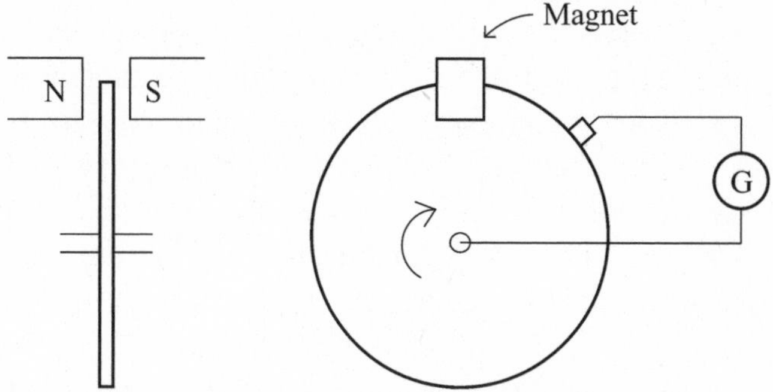

Fig. 5.5. Schematic layout of Faraday's first dynamo. (Used with permission from Lee Bartrop.)

By the same token, he had explained Arago's result: When Arago spun his disc, its movement relative to the suspended magnet had produced a current in the disc that, in turn, set up magnetic lines of force that caused the magnet to rotate.

As he had done with the electric motor, Faraday was content to have demonstrated the principle of the dynamo and to leave its technological development to others. His job was to try to push forward the frontiers of knowledge about the physical world. He had made startling new discoveries about the way electricity and magnetism behaved, and all his thoughts were now absorbed in trying to explain them.

In fact, he had been thinking hard on the topic for eleven years already. Ever since hearing of Oersted's discovery of the magnetic needle that turned perpendicular to the current, he had been trying to understand the relationship between electricity and magnetism. His ideas came from what he had observed for himself in the laboratory; they owed nothing to mathematics or to the theories of others. Vague and tentative at first, they were now slowly beginning to coalesce, and they were not like anything that had come before. Faraday knew that many orthodox scientists would dismiss his ideas as laughable, or even heretical, and so kept them largely to himself.

On November 24, 1831, he presented his discovery of electromagnetic induction to his colleagues at the Royal Society. The paper was sparsely worded and factual, reporting experimental results with no fanfare and almost no conjecturing. Not yet ready to mount an open challenge to established theories, he chose his wording carefully, invoking only the part of Ampère's model that postulated a connection between electricity and magnetism. He ventured into theoretical territory only briefly, but here he gave just a glimpse of how far away he was from Ampère, or from anyone else. Reporting his experiments with the iron ring, he explained how he believed the current was induced in the secondary circuit:

> Whilst the wire is subject to either volta-electric or magneto-electric induction, it appears to be in a peculiar state; for it resists the formation of an electrical current in it, whereas, if in its common condition, such a current would be produced; and when left uninfluenced it has the power of originating a current, a power which the wire does not possess under common circumstances. This electrical condition of matter has not been recognized, but it probably exerts a very important influence in many if not most phenomena produced by currents of electricity. For reasons which will immediately appear, I have, after advising with several learned friends, ventured to designate it as the electro-tonic state.[2]

Perhaps it is not surprising that few, if any, of Faraday's colleagues had the faintest idea what he was talking about. As we'll see, the electrotonic state went on to play a big part in the development of field theory, but at this stage it was an enigma, even to Faraday. It appeared to be a kind of tension or strain that was present in a wire whenever the wire carried a current, but it revealed itself only while it was being set up and while it was being released—he tried everything he could think of to detect the strain directly, but without success. Faraday's conception of this elusive state was an extraordinary feat of scientific imagination. The most puzzling aspect of the iron-ring experiment was the brief current that flowed in the secondary circuit *after* the battery in the primary circuit had been disconnected. To Faraday, this seemed to be the release of some kind of strain in the secondary circuit that had been induced by the primary current through the agency of the magnetic force in the iron ring that linked the two circuits. The brief current that had flowed *in the reverse direction* in the secondary after the battery was first connected in the primary circuit was, by the same token, a manifestation of the *setting up* of this strain. The state of strain continued while the primary current was steady but collapsed when the battery was disconnected, and the brief secondary current then flowed rather in the way that gas is released from a pricked balloon.

Such thoughts sprang from well-prepared soil. Faraday had been thinking deeply along lines different from everyone else's. An electric current might, he imagined, be the result of a rapid buildup and release of a strain in a wire. And every electric current created a circular magnetic force around itself—a force that seemed to have a physical presence in the surrounding space. Similarly, every magnet created a pattern of forces that acted along *curved* paths in the space around it. Though invisible, these paths needed only a sprinkling of iron filings on a piece of paper held over the magnet to reveal themselves.

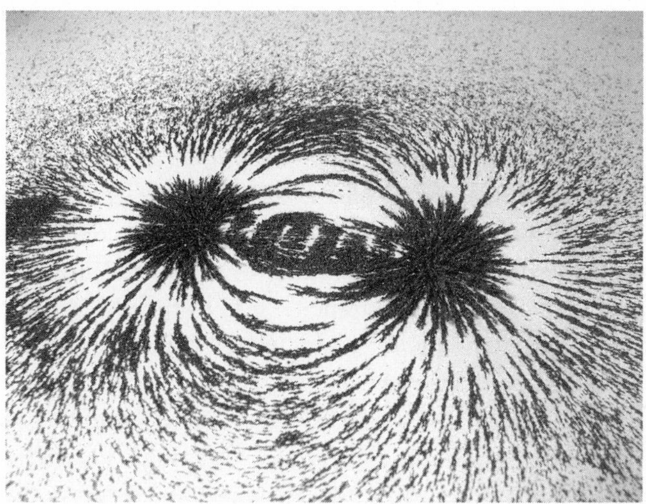

Fig. 5.6. Magnetic lines of force shown by iron filings sprinkled on paper over a magnet. (Courtesy of Windell H. Oksay / www.evil madscientist.com.)

The iron filings showed a cross section of a three-dimensional pattern that Faraday saw in his mind's eye, not only around every iron magnet but also around every current-carrying wire coil. (Such coils acted as magnets, as Ampère had shown.)

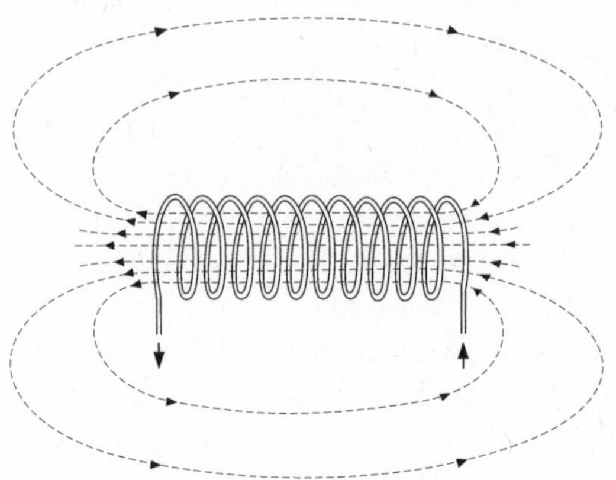

Fig. 5.7. Magnetic lines of force around a current-carrying coil. (Used with permission from Lee Bartrop.)

The pattern changed when a second magnet was brought in, and one of its poles placed near a pole of the first magnet. He explained the notion in his *Experimental Researches in Electricity*: "By magnetic curves, I mean the lines of magnetic forces, however modified by the juxtaposition of poles, which would be depicted by iron filings."[3]

This was the first public appearance of the term "lines of force"—a concept central to the development of field theory. Along with the electrotonic state, it helped explain what he had discovered in the laboratory. A current was created in a wire circuit when the wire moved near a magnet, or vice versa; that is, when the wire *cut across* the magnet's lines of force. This act of "cutting" magnetic lines of force, Faraday surmised, was what produced an electrical force in the wire. If the wire was part of a complete circuit, a current then flowed. For example, by pushing a magnet into the cavity of a coil, Faraday had caused its lines of force to be "cut" (in the sense of crossed, not severed) by all the turns of wire in the coil, and thus he generated a current. The same happened in reverse when he pulled the magnet away, and a current flowed in the other direction.

What about his dynamo? At any instant, a part of the copper disc was moving between the poles of the magnet and cutting its lines of force, hence the steady current across the disc. The same principle held in the iron-ring experiment, though in a less obvious way. When the primary current was first switched on, a wave of lines of magnetic force spread out from the primary coil around the iron ring and into the surrounding space, and as they spread through the secondary coil, they were cut by all the turns of wire there, causing a current to flow. Once the lines of force had settled to a fixed pattern, they were no longer being cut and the secondary current stopped. But when the primary current was turned off, the process was reversed; the lines of force were cut again while they receded, and a current flowed in the opposite direction in the secondary circuit.

By his unique combination of faculties, complementing supreme experimental skill with sustained imaginative thought, Faraday had produced two entirely new concepts: (1) the electrotonic state and (2) lines of magnetic force. He felt sure they both fit into a larger picture, but at this stage the picture was still a hazy one. Meanwhile, he had an unwelcome taste of déjà vu.

As we have seen, Faraday had reported his discovery of electromagnetic induction to the Royal Society on November 24, 1831. Around this time, he had also sent word of it to Paris, and his correspondent, J. N. P. Hachette, read out his letter at the December meeting of the Académie des Sciences. So far, so good, but a garbled account of the meeting got to the magazine *Le Lycée*, which published it and followed up with an article that attributed the discovery to two French scientists. Meanwhile, two Italians, Leopoldo Nobili and Vincenzo Antinori, who had seen the first *Lycée* article, repeated the experiments and published their own findings. They acknowledged Faraday but, by some mischance, the journal in which their article appeared carried the false date November 1831, whereas Faraday's paper didn't appear in print until early 1832. Back in London, things went from bad to worse: William Jerden of the *Literary Gazette* spotted the difference in dates and told his readers that two Italians had just beaten Michael Faraday to a great discovery.

A decade after being falsely accused of plagiarizing Wollaston's ideas on electromagnetic rotations, Faraday was once again innocently embroiled in controversy. All his Sandemanian self-control couldn't contain his fury. He wrote to Jerden:

> I never took more pains to be quite independent of other persons than in the present investigation; and I have never been more annoyed about any paper than the present by the variety of circumstances which have arisen seeming to imply that I had been anticipated.[4]

Hachette and Jerden made private and public apologies; and the English translation of Nobili's and Antinori's article carried a prominent acknowledgement of Faraday's prior claim. But this was the last time Faraday freely announced his results before publication.

His distress from this episode was balanced by the joy of having discovered, by simple experiment and plain reasoning, facts that had eluded all the mathematical physicists. His Sandemanian conscience was put to the test again. This time he succumbed, if only for an instant, to the sin of pride, writing to Richard Phillips:

> It is quite comfortable to me to find that experiment need not quail before mathematics but is quite competent to rival it in discovery and I am amused to find that what high mathematicians have announced . . . has so little foundation. . . . Excuse this egotistical letter.[5]

Faraday's thoughts on electricity and magnetism ran on—his conjecturing stretched the imagination but was always based on what he had observed in experiments. While carrying out his experiments on acoustic vibrations and waves, he had begun to wonder whether he was watching a relatively slow-motion version of what happened with electricity and magnetism, and his new results with magnets and electric circuits had hardened that thought almost into a conviction. The establishment of magnetic lines of force around the wire when a battery was connected to its circuit must, he thought, take time, and magnetic and electric forces must be transmitted over time by vibrations or waves in the intervening medium, rather like acoustic pressure in air. Such thoughts ran completely counter to the prevailing theory of instantaneous action at a distance, but he had no direct evidence to support them and so held back from mentioning them in his published papers. There was, however, a way to register these radical thoughts formally while still keeping them private. In March 1832, he asked the secretary of the Royal Society to deposit a note in his safe. It read:

> Certain of the results embodied in the two papers Experimental Researches in Electricity and Magnetism . . . led me to believe that magnetic action is progressive and requires time, i.e. that when a magnet acts on a distant magnet . . . the influencing cause . . . proceeds gradually from the magnetic bodies and requires time for its transmission which will probably be found to be very sensible. I think also that electrical induction (of tension) is also performed in a similarly progressive time.
>
> I am inclined to compare the diffusion of magnetic forces from a magnetic pole to the vibrations upon the surface of disturbed water or those of air in the phenomenon of sound; i.e. I am inclined to think the vibratory theory will apply to these phenomena, as it does to sound and, most probably, to light.[6]

Having twice been wrongly accused of plagiarism, Faraday was probably taking a precaution against the same thing happening again.

There were now several known sources of electricity. Some fish, like electric eels, generated their own; friction produced static electricity, which could be stored in devices like the Leyden jar that released all their charge in one burst; voltaic batteries produced a steady electric current from chemical action between two different metals; and there were Faraday's new magneto-electric currents. Were all the types of electricity identical? Wollaston and others had shown that static and voltaic electricity produced similar electrochemical effects, but, as always, Faraday had to see things for himself, and now he decided to carry out a thorough investigation. He first identified six different types of electrical effect: the attraction and repulsion of electric charges; the heating effect of a current; the production of magnetic forces; chemical decomposition; physiological effects; and, lastly, the spark. He then worked through these effects systematically and demonstrated to his own satisfaction that they were the same whatever the source of the electricity, not neglecting electric fish. As to what electricity actually *was*—one fluid, two fluids, or something else—he kept an open mind. This agnosticism is evident from the description of an electric current that he gave in his *Experimental Researches in Electricity* in 1833.

> By current, I mean anything progressive, whether it be a fluid of electricity, or two fluids moving in opposite directions, or merely vibrations, or, speaking still more generally, progressive forces.[7]

Currents didn't exist just in wires; they also flowed in chemical solutions, and there, Faraday thought, he might discover more about them. In the course of yet another historic series of experiments, he established the two fundamental laws of electrolysis: the mass of a substance in a chemical solution that is decomposed when a current is passed through it is proportional to the total amount of electricity passed; and the masses of different substances produced by a given amount of electricity are proportional to what is called their "equivalent masses." The equivalent mass of an element is now defined in terms of its atomic structure. We can only wonder at the genius of a man who was able to establish this law seventy years before the existence of atoms was proved, especially since Faraday held similar agnostic views about atoms to those that he held about electricity. He wrote:

> The equivalent weights of bodies are simply those quantities of them which contain equal quantities of electricity, or have naturally equal electric powers; it being the ELECTRICITY which determines the equivalent number, because it determines the combining force. Or, if we adopt the atomic theory or phraseology, then the atoms of bodies which are equivalents to each other in their ordinary chem-

> ical action, have equal quantities of electricity naturally associated with them. But I must confess I am jealous of the term atom, for though it is very easy to talk of atoms it is very difficult to form a clear idea of their nature, especially when compound bodies are under consideration.[8]

When giving a lecture in memory of Faraday almost half a century later, the great German physicist Hermann von Helmholtz shone a light on Faraday's extraordinary prescience. He said:

> The most startling result of Faraday's law is perhaps this. If we accept the hypothesis that the elementary substances are composed of atoms, we cannot avoid concluding that electricity also, positive as well as negative, is divided into elementary portions, which behave like atoms of electricity.[9]

The fact that Helmholtz was speaking a decade before the electron was discovered makes his words still more potent. Today we may wonder why Faraday didn't pursue the possibility that each atom has a positively charged part and a negatively charged one. A clue lies in his strange use of the word *jealous* to describe his opinion of atoms. Like his mentor Davy, Faraday distrusted John Dalton's theory that all matter was composed of atoms, even though it offered a simple explanation for the proportional weights of elements in chemical reactions. Both Davy and Faraday sought unifying theories and didn't like the way that Dalton had apparently divided chemical substances into many unrelated types, each with its own kind of atom. Yet Dalton's simple theory turned out to be fundamentally correct—in this rare case we might say that Faraday threw the baby out with the bath water.

Meanwhile, Faraday faced the question, how does a current flow in a chemical solution? The prevailing view was analogous to the action-at-a-distance theories of forces between charged objects and between magnetic poles: When the ends of wires connected to a battery were dipped in the liquid, the wire ends became centers of forces that acted along a straight line between them and tore apart the particles of the substance in the solution. The two parts of each torn-apart particle were oppositely charged, so one was drawn toward the negative wire end; the other was drawn to the positive; and the movement of all the free fragments, constituted the current. So prominent was this view that the wire ends were called poles, by analogy with magnetic poles.

Faraday's mentor Davy, along with Theodor Grothuss from Leipzig, thought the process was rather more complex. In their interpretation, forces from the positive and negative poles did not simply tear the particles apart; they set up a chain of chemical exchanges in the solution, in the course of which the charged fragments continually changed partners in such a way that the positively charged parts moved one step at a time toward the negative pole and the negatively charged ones moved the other way—rather like buckets in a two-way human bucket chain. When the positively or negatively charged fragments reached their respective wire ends, they forsook their last temporary partners, donated their charges to the wire, and emerged free.[10]

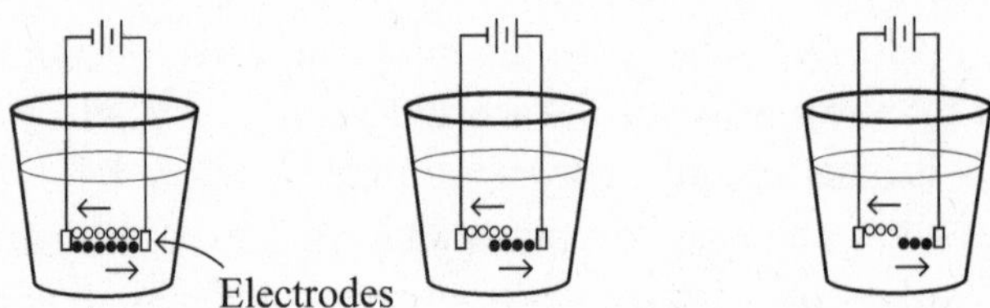

Fig. 5.8. Moving chain of charged particles in electrolysis. (Used with permission from Lee Bartrop.)

In one of his brilliant flashes of insight, Davy had taken a further step by saying that electricity was the force that bound different elements together into compounds, implying that it was an inherent property of matter. No need for two fluids, or even one. He was, of course, right, but the notion of fluids had taken such a firm hold that many years passed before they were finally abandoned.

Faraday sided with Davy and Grothuss but went still further by refuting the role of poles altogether. Whereas they thought that it was the forces from the poles that caused the chemical exchanges, and that these forces diminished with distance, Faraday believed that the poles exerted no force at all—they were simply the entry and exit ports for the electric current that flowed through the solution, powered by the battery. He was right.

When writing up his researches on electrochemistry, Faraday was faced with a difficulty: he was trying to describe physical processes that nobody had described, or perhaps even thought of, before. This was not a new problem; he had already, with the help of friends, invented the term *electrotonic state*. As his ideas grew, language itself became part of his thinking. Each new word and each new phrase helped to clarify, even define, the underlying concept, even if that itself was not fully formed. He sought precision and faced a particular problem with words that were in common use but had misleading theoretical connotations. The prime example was *current*, which implied that electricity was a fluid. Here he never found a useful alternative—nor has anybody since—but elsewhere he coined new words that eventually became the standard ones. By careful and creative attention to language in his writings, he often made it possible for others to grasp and accept new ideas, even when they ran completely counter to prevailing theories. There were exceptions; some of his concepts were so different from anything seen before that none of his contemporaries understood them, especially as he was unable to express them in mathematical terms. His successor at the Royal Institution, John Tyndall, explained the difficulty:

> It sometimes strikes me that Faraday saw the play of fluids and ethers and atoms though his previous training did not let him resolve what he saw into constituents or describe it in a way satisfactory to a mind versed in mechanics. . . . It must, however, always be remembered that he works at the very boundaries of knowledge and that his mind habitually dwells in the "boundless contiguity of shade" by which that knowledge is surrounded.[11]

As we'll see, it took James Clerk Maxwell to understand how Faraday "saw the play" and to translate Faraday's ideas into the kind of mathematical language that others could understand.

When he needed a new word, Faraday took care to draw on the best advice. He now consulted his friend Dr. Whitlock Nicholl. With Nicholl's help, Faraday now proposed to do away with the misleading term *pole* for the circuit terminals that dipped into the solution and to replace it with *electrode*. Hence, also, came *electrolysis* for the process of separating the components of a solution by passing an electric current through it, and *electrolyte* for the solution so treated. For further advice, Faraday turned to the Cambridge polymath William Whewell, who proposed *anode* for the positive electrode, *cathode* for the negative one, *anion*, *cation*, and more generally *ion*. Like many innovations, the new words met sharp resistance at first, but Faraday was, by now, worldly wise and he countered it by making full use of Whewell's formidable reputation. He acknowledged the debt in a letter to Whewell:

> I had some hot objections made to them here and found myself very much in the condition of the man with his son and Ass, who tried to please every body; but when I held up the shield of your authority it was wonderful to observe how the tone of objection melted away.[12]

Remarkably, all the common terms now used in electrolysis, except for the stubborn old survivor, *current*, are the ones that Faraday created in the 1830s with the help of his learned friends. They are testament to the immense care he always took to describe his findings in accurate language, and to his skill in communication. New words were needed not for their own sake but because existing ones carried theoretical baggage that could constrain one's thinking. He may not have known classical Greek, but he had the *savoir faire* to consult scholars who did, and to choose the right words when they were offered—words that conveyed their meanings clearly, memorably, and with no prior theoretical connotations.

From his early days with Davy, Faraday had been exposed to the idea that electricity was an inherent power of matter. Now he knew it for a fact. His investigations had also shown beyond doubt that, in electrochemistry at least, electrical force was transmitted not at a distance but locally from particle to particle. Moreover, the force did not act in straight lines but in curves. For example, when two wire terminals from a battery were dipped in a solution of copper chloride, copper was deposited not just on the side of the cathode that directly faced the anode but all around; if the cathode was in the shape of a blade, it became copper-plated at the back as well as the front, showing that the chains of chemical exchanges, and hence the forces, must follow curved paths.

These were remarkable results, but mathematical physicists, whether in Britain, in France, or elsewhere paid them little attention. Messy, smelly chemistry was outside their province, and they cared little for it. Things were different, however, when Faraday ventured into their heartland of electrostatics, where Coulomb's law, with all its connotations of straight-line action at a distance, was sacrosanct. How dare a mathematical illiterate like Faraday poke his nose in their domain?

He was prompted to turn to electrostatics by a question that arose naturally during his investigations in electrochemistry. How does the process of electrolysis start? What happens when the battery is first connected to the circuit? Faraday reasoned that immediately before decomposition begins, all the particles in the chemical solution, or electrolyte, must, if only for a moment, be in a polarized state—that is, in a state of tensile strain—with their positively and negatively charged parts being pulled in opposite directions. And, while in this state, they will tend to arrange themselves in chains, each particle held close to its neighbors in a sequence of alternating positive and negative charges. These, he surmised, are the chains along which the chemical exchanges will take place—curved lines of electrical action. This state of strain is maintained only fleetingly in the electrolyte—as soon as the particles begin to decompose, the strain is released, at least in part, and chemical exchanges take place along the curved lines of electrical action, constituting a current.

For Faraday this was more than just an explanation of how chemical decomposition begins in an electrolyte, it was the starting point of a process of thought that completely upended the Newtonian view of the physical world. A grand scheme began to form in his mind. Perhaps the same polarized state of strain that starts the process of electrolysis was actually the source of *all* electrical action in materials. If the electrolyte were replaced with a metal conductor, the metal would not be able to support the strain in a sustained way and a current would flow freely. On the other hand, if the electrolyte were replaced by an insulating material, the strain would be maintained—though under extreme strain the insulation would eventually break down and there would be a sudden discharge, like a spark through air. In all likelihood, Faraday thought, no materials were actually perfect conductors or perfect insulators; even the best conductors would hold onto a small part of the strain, and even the best insulators would leak a little current. But the idea that insulators could maintain the polarized state of strain, even if not quite perfectly, had momentous implications. This state of strain in insulators seemed to Faraday to be an electrical equivalent of the electrotonic state that he had supposed to exist in and around a wire when it was close to a magnet. Like the electromagnetic type of electrotonic state, the electrostatic one proved impossible to detect directly, though Faraday tried everything he knew to try to reveal it.

There were two principal electrostatic effects that traditional theories attributed to action at a distance. One was the mechanical force between charged bodies. The other was electrical induction: a charged body induced an opposite charge in another body nearby. Faraday could now explain how this happened. There was no mysterious action at a distance; induction occurred along the chains of contiguous polarized particles running from one body to the other through the insulating medium between them. If the first body were positively charged, it would attract the negative parts of the particles of the intervening medium that were next to its surface, so the particles at the far end of the chains next to the surface of the second body would be positive and would attract negatively charged parts of particles on the surface of that body, thus inducing a negative charge there.

If Faraday's idea were correct, the inductive effect between the initially charged body and the other would depend on the propensity of the intervening medium to form the inductive chains, and so it would be expected to vary from one type of insulating substance to another. He made an apparatus to test this very point, using two concentric metal spheres with various substances, including air, shellac, wax, and sulfur, in the space between them, and found that the effect did, indeed, vary widely. Each substance had its own specific inductive capacity—another great discovery. In the course of this experiment, Faraday was able to observe something he already suspected: Induction took time to act through the insulating material—a finding that later posed a problem for the Atlantic telegraph-cable project.

What about the *forces* between charged bodies? While imagining the paths that the charged fragments of particles followed when traveling between the electrodes in a tank of a chemical solution, Faraday had been reminded of the pattern that iron filings took when sprinkled on a piece of paper over a magnet. It was this pattern that had prompted his idea of lines of magnetic force to explain electromagnetic induction. Now he had envisaged chains of contiguous polarized particles to explain *electrostatic* induction, and the pattern of curves they formed in his mind also resembled that of iron filings around a magnet. These chains of induction were, surely, lines of *electric* force. The tension along chains of stretched polarized particles would explain why opposite charges attracted one another.

Faraday also surmised that his lines of electric and magnetic force repelled one another sideways, though the full significance of this property seemed to come to him only slowly—an eminent biographer reports that he was "dimly aware" of it at this time.[13] This seems curious, as the repulsion between lines of force so neatly explains that between like charges and between like poles.[14] It is not hard to find other examples of Faraday's fallibility. He had earlier toyed with the (correct) idea that the electromagnetic version of his electrotonic state bore a close analogy to momentum in mechanical systems, only to reject it.

In fact, these examples serve to remind us that he was working in completely unknown territory, struggling to make sense of the strange, and sometimes apparently contradictory, findings from his experiments. The wonder is not that he missed the odd point but that he somehow managed, from such confusing evidence, to produce ideas that were so unusual as to be almost impossible to describe in words, yet that turned out to be correct. Some of his ideas were not understood by anyone else until first the great physicist William Thomson (Lord Kelvin) and then James Clerk Maxwell, both Scotsmen from the following generation, expressed them in mathematical language.

By Faraday's scheme, the electric charges that appeared on the objects were simply the end points of lines of induction, which had to be positive at one end and negative at the other so the total net charge was always zero. And because conducting substances like metals cannot support inductive strain, the charge exists only on their surfaces, where they abut the lines of induction. One consequence of these supposed properties of matter was that no induction from outside would penetrate a closed metal container. Faraday carried out many experiments on this and related themes, but the most spectacular one was performed for an astonished audience in the Royal Institution lecture theater. Faraday, the showman, built a wooden-framed cube twelve feet across, coated it with tin foil, and, with the audience present, stepped inside. The metal surface was then charged by an electrostatic generator to many thousand volts. Sparks flew from the corners while Faraday, calmly sitting inside, checked that none of the charge had penetrated inside the box. This was the first demonstration of the Faraday cage. Now everyone travels in cars and airplanes confident that if lightning strikes the vehicle, no harm will come to the passengers.

He chose a simple experiment to show that electrostatic induction does not always act in straight lines, as the action-at-a-distance adherents believed, but can act along curved paths. Faraday placed a brass ball near a negatively charged shellacked rod. Metals did not transmit electrostatic induction, so if the induction acted in straight lines, the ball would have acted as a screen and cast a sharply defined shadow—a region which no induction from the rod would reach. But he found that the negative charge on the rod actually induced a positive charge on objects placed entirely within the supposed shadow. The lines of induction must have curved around the brass-ball screen. This, to Faraday, was confirmation not only that electrostatic induction, and hence electrostatic forces, acted along curved lines but also that they acted from particle to particle—since that was the only way they could follow curves.

By June 1838, he was able to include in his published *Experimental Researches in Electricity* a ten-point summary of his theory of the nature of static electricity:

> The theory assumes that all the particles, whether of insulating or conducting matter, are as whole conductors.
>
> That not being polar in their normal state, they can become so by the influence of neighbouring charged particles, the polar state being developed at the instant, exactly as in an insulating conducting mass consisting of many particles.
>
> That the particles when polarized are in a forced state, and tend to return to their normal or natural condition.
>
> That being as whole conductors, they can readily be charged, either bodily or polarly.
>
> That particles which being contiguous are also in the line of inductive action can communicate their polar forces one to another more or less readily.
>
> That those doing so less readily require the polar force to be raised to a higher degree before this transference or communication takes place.
>
> That the ready communication of forces between contiguous particles constitutes conduction, and the difficult communication insulation. . . .
>
> That ordinary induction is the effect resulting from the action of matter charged with excited or free electricity upon insulating matter, tending to produce in it an equal amount of the contrary state.
>
> That it can do this only by polarizing the particles contiguous to it, which perform the same office to the next, and these again to those beyond; and that thus the action is propagated from the excited body to the next conducting mass, and there renders the contrary force evident in consequence of the effect of communication which supervenes in the conducting mass upon the polarization of the particles of that body.
>
> That therefore induction can only take place through or across insulators; that induction is insulation, it being the necessary conse-

> quence of the state of the particles and the mode in which the influence of electrical forces is transferred or transmitted across such insulating media.[15]

In brief, static electricity showed itself in material substances as a form of strain that was passed on from each particle to its neighbors. In a closed circuit, when the strain collapsed, a current flowed. Faraday was writing at a time when most physicists still believed that electricity was an imponderable fluid (or two) that exerted forces at a distance along imaginary straight lines. One of his biographers, Sylvanus P. Thompson, has aptly described how he was able to see far beyond the range of his contemporaries:

> Living, working and daily investigating in his laboratory, in the presence of all his apparatus, he gave his thoughts free play around the phenomena, incessantly framing theories to account for the observed facts and then testing his ideas by experiment, never hesitating to push these ideas suggested by his experiments to their logical conclusion, no matter how much they may have diverged from the accepted scientific theories of the day. . . . [H]e worked on and on with a scientific foresight which could be called miraculous. His experiments, even those which at the time seemed unsuccessful, in that they yielded no immediate positive results, have proved to be a deep mine of richness for the scientific minds that followed him.[16]

Faraday's accomplishments during the 1830s almost defy belief. By 1838, his set of *Experimental Researches in Electricity* had reached series 14. And beside his duties as director of the Royal Institution, he also lectured at the Royal Military Academy in Woolwich and, of course, in the Institution's own theater. He also broke his earlier resolution by taking on some important analysis and consultancy jobs, some from a strong sense of social or patriotic duty, and some to top up the institution's coffers or to boost his modest salary. One of these was to act as scientific adviser to Trinity House, the organization responsible for Britain's lighthouses. Sarah took him away to the country or the seaside when she could, but the work began to take its toll. The nervous headaches that had occasionally afflicted him in his youth became more frequent; he could no longer keep up the intense concentration that his style of work required; and he suffered increasingly from lapses of memory.

The doctor ordered a month off, then another and another. Faraday went to the laboratory only intermittently and found that even writing to friends was hard work. In one letter, he said his memory was so treacherous that by the time he got to the middle of a sentence he couldn't remember how it started. On his forty-ninth birthday, Faraday felt sure that his days of making great discoveries were over. He was wrong.

CHAPTER SIX

A SHADOW OF A SPECULATION

1840–1857

Faraday, the compulsive experimenter, did no work in his laboratory for two years. Refreshed by enforced holidays at the seaside and in the mountains, he glowed with physical health, thinking nothing of walking thirty miles in a day over an Alpine pass. He took delight in his family and friends and in the beauty and power of nature—he loved to run out in a thunderstorm—but without experimental work, his life was incomplete. He summed things up in a notebook entry, "I would gladly give half my strength for as much memory—but what have I to do with that? Be thankful."[1]

Strictly rationing his mental exertion, Faraday did what work he could. He gave lectures, including the Christmas ones to children in 1841 and 1842; advised Trinity House on ventilation in lighthouses; and led government inquiries into explosions at a gunpowder factory and a coal mine. In 1842, he was tempted to return to the laboratory to investigate William Armstrong's well-publicized discovery that steam issuing from a boiler was electrically charged. After finding that the charge came from frictional contact between water droplets and the vent pipe, he again left the laboratory to Sergeant Anderson's care, not returning until 1844, when he resumed some earlier work on the liquefaction of gases and made a little progress, succeeding with ammonia and with Davy's one-time specialty, the mind-altering nitrous oxide.

In June 1845, Faraday went to the annual meeting of the British Association for the Advancement of Science. He wasn't a regular participant, but this year's gathering was at Cambridge, where William Whewell, who had suggested the words *ion*, *anode*, and *cathode* to him, was master of Trinity College. Between sessions, an engaging young Scot introduced himself—he was a newly elected fellow of Peterhouse, a college at Cambridge—by the name of William Thomson. Now remembered as Lord Kelvin, Thomson was a prodigy who had enrolled at Glasgow University at the age of ten and taken top prizes in all subjects. Mathematics was a particular passion of his, and at the age of seventeen he had shown that Faraday's electric lines of force could be represented by the same equations that Joseph Fourier had derived for the flow of heat in a metal bar. This was a historically significant paper, giving the first indication that Faraday and mathematics were compatible, but, coming from a teenager in his first term at Cambridge, it had attracted little attention. Faraday seems not to have noticed it. All the same, he cannot have failed to be impressed by the young man with the quicksilver mind—everybody was—and when a follow-up letter from Thomson arrived in August, he was inspired to take up the path of discovery once more.

To Thomson, as to Ampère, mathematics was the language of science. Where Faraday had to do his own experiments to understand a topic, Thomson had to write his own equations. His first impression of Faraday's *Experimental Researches in Electricity*, which gave not a single equation, was that they seemed to be written in a perversely cumbersome foreign tongue, but once Thomson saw the analogy of lines of force with Fourier's mathematical theory of heat flow, he began to take the idea of lines of force seriously—the first person, apart from Faraday himself, to do so. He was intrigued to discover that exactly the same results could be derived from Coulomb's and Ampère's theory of electrostatic forces, so were Faraday's lines of force simply another way of formulating instantaneous action at a distance between point charges? Thomson thought so at first, but he noted Faraday's finding that electrical induction took time to act, rather like Fourier's heat flow; and he found that the using the analogy between lines of force and heat flow actually made some calculations a lot simpler. He began to think that Faraday could be right—that lines of force could have a physical existence and that electrical forces could be the manifestation of some kind of strain in the medium between the charged objects.

Thomson knew that internal strains in a transparent substance could be detected by shining polarized light through it—that is, light in which the transverse wave vibrations are all lined up in a particular plane rather than being randomly oriented as in ordinary sunlight. Scientists had found that when polarized light passes through a mechanically stressed transparent substance, the alignment of its vibrations, formally the plane of polarization, is altered. Taking Faraday's idea that electric lines of force represented a kind of strain in the medium that carried them, Thomson wondered whether that strain might be detected by subjecting a transparent substance to electrical force, shining a beam of polarized light through it, and observing any change in the plane of polarization. Perhaps Faraday had already tried the experiment; Thomson wrote to ask.

Over the years, Faraday had been plagued by futile suggestions from well-meaning enthusiasts, but this one was different. He had indeed tried the experiment that Thomson had suggested—more than once and to no effect—but it was a good suggestion, and, amazingly, it had come from a young mathematician who had taken the trouble to try to understand his work. This was just the stimulus Faraday needed; he thanked Thomson promptly and resolved to try again.

He first tried sending the polarized light through various liquids undergoing static electrification from an electrostatic generator: distilled water, sulfuric acid, and solutions of copper sulfate and sodium sulfate. He shone the light beam first parallel to the direction of electrification and then across it, but the direction of its transverse vibrations remained unmoved. Undeterred, he tried the effect of electric currents—steady, rising, falling, pulsed—and, finally, sparks. Then he substituted various solid, transparent substances for the liquids—plate glass, quartz, Iceland spar, and others—but he found that he had nothing to show for a fortnight's work.

Perhaps the change in the plane of polarization had been too small to detect. He decided to try the effect of magnetic forces, which could be made much stronger than electric ones—he had powerful electromagnets in his Royal Institution laboratory and could call on an even bigger one from the Royal Military Academy in Woolwich. Once again, he varied the conditions of the experiment, changing the position and strength of the magnet and shining polarized light through all the transparent substances again, all to no avail. Then he thought of the small sample of special heavy glass, made of borosilicate of lead, left over from his onerous work on the glass project of the 1820s.

For two weeks he had been looking into a lens and seeing nothing but black. The source of light was an oil lamp, and the light passed first through a polarizing prism, then through the test substance, and finally through an analyzer, before arriving at the eyepiece. The apparatus was arranged so that only if the light's plane of polarization were *altered* by the action of electric or magnetic force on the test substance would any light reach the eyepiece. At first, he put the piece of heavy glass between the north pole of one magnet and the south pole of another. Nothing happened. Nor did it when he tried several other configurations. But when he placed the piece of glass *alongside* the same two poles, shone the beam, and looked into the eyepiece—there was a faint image of a flickering flame.

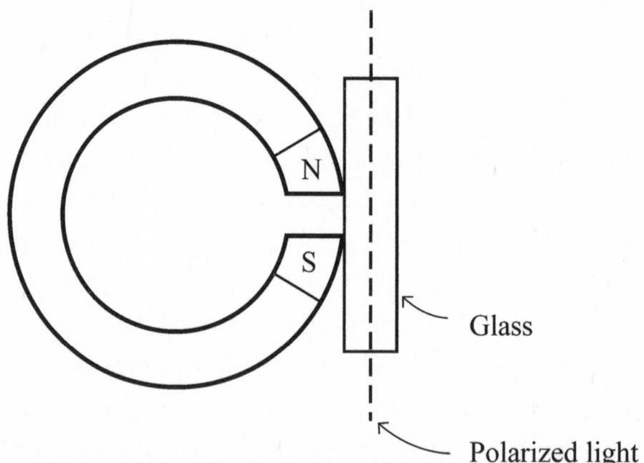

Fig. 6.1. Arrangement of magnet and heavy glass in Faraday's magneto-optic experiment. (Used with permission from Lee Bartrop.)

Faraday once more scented discovery. He had opened the door a crack but needed more powerful equipment. That was easily arranged; he sent for the great Woolwich electromagnet and took a four-day break in pleasant anticipation, planning what he would do when it arrived.

On September 18, he repeated the experiment with the sample of heavy glass, this time using the Woolwich magnet. The image of the flame was there, much brighter than before, but it took a second or so to rise to full brightness. This was the best demonstration yet of something Faraday had observed before—electromagnets take time to reach their full intensity of magnetization. He then verified that the light's plane of polarization did indeed rotate as it traveled through the magnetized glass, determined the direction of rotation, and found that the angle of twist was proportional to the strength of the magnet (which could be altered by varying the number of turns of wire coiled around the iron that was connected to the battery). Working with all the energy of his youthful days, he carried out test after test, noting the results in his laboratory log along with such expressions as "very fine effect" and "effect was best yet." After filling twelve pages, he closed the log with the words, "An excellent day's work."[2]

With the big Woolwich magnet at hand, Faraday now retested the various transparent substances that had earlier failed to produce any discernible rotation in the light's plane of polarization. They all now did so, and Faraday felt that his belief in the deep-lying unity of all nature's forces was being borne out. He had shown not only that light and magnetism were in some way connected but also that glass and a variety of other transparent substances, hitherto believed to be nonmagnetic, were affected by magnetism. The results prompted the question, are *all* substances in some way magnetic? This wasn't a new thought; he had actually already tried the effect of a magnet on many substances with no result. But now the muse was with him and he resolved to have another go.

The most direct demonstration of magnetism in a supposedly nonmagnetic substance would be to put a sample of it between the poles of a strong magnet and get it to move, like a compass needle. The most promising substance was his piece of heavy glass, which was in the shape of a bar that was two inches long and half an inch square, but when he put it in a small paper sling and suspended it by a silk thread between the poles of the Woolwich magnet, there was no discernible effect. This disappointment didn't deflect Faraday from his task. Perhaps an even stronger magnet was needed; he procured half of a huge iron link from a ship's anchor chain and had it made it into a gigantic electromagnet, wound with 522 feet of wire and weighing, in all, 238 pounds.[3] While this monster was in preparation, he tried many more experiments with the equipment he already had in the laboratory. All were designed to find further evidence of connections between nature's forces and all gave negative results, but in one experiment he narrowly missed detecting what later became known as the Kerr effect, in which the polarization of light is modified by reflection from a magnetized metal surface.

On November 4, the new magnet was ready. When Faraday suspended his bar of heavy glass between its north and south poles, the bar swung and finally aligned itself *at right angles* to the lines of magnetic force. He had shown that glass had a new kind of magnetic property—one that didn't depend on light. The door was open to yet another great discovery. He substituted many other supposedly nonmagnetic substances for the glass, and all behaved in the same way. Crystals, powders, and various liquids (in thin containers), wood, beef, apple, bread, and even most metals aligned themselves at right angles to the magnetic lines of force. Iron, cobalt, and nickel were the exceptions that aligned themselves *parallel* to the lines of force. Summarizing the results vividly in his *Experimental Researches in Electricity*, Faraday wrote:

> If a man could be suspended with sufficient delicacy . . . and placed in the magnetic field, he would point equatorially, for all the substances of which he is formed, including the blood, possess this property.[4]

Just as significant, in its way, as the great new discovery is Faraday's use here of the term *magnetic field*, which made its first appearance in his notebook a little earlier. The idea that space itself could be the seat of forces was now encapsulated in a word that would become indispensable to physicists—the field.

Faraday was so taken up with the new work that he didn't

want to leave the laboratory to attend the Royal Society meeting on November 20, at which his paper "On the Action of Magnets by Light" was to be presented—someone else had to read it for him. His spectacular results drew him to conclude that all solid and liquid substances react to magnetic forces—probably gases, too, though for the present he couldn't think how to prove it. He found that the substances that aligned themselves across the magnetic field, rather than parallel to it, were repelled by *any* magnetic pole, whether north or south. The great majority of substances were in this category, and he needed a new word to describe them. With William Whewell's help, he chose *diamagnetic*. A new word was now also needed to describe the minority of substances that aligned themselves parallel to the field—they had been called simply *magnetic* but that would no longer do now that *all* substances were believed to have magnetic properties. The name Faraday chose for materials such as iron, nickel, and cobalt was *paramagnetic*.

Heartily encouraged by these successes, Faraday went on with his search for ways of unifying nature's forces. He now believed that magnetism was a universal property of matter, and he knew that it could affect a ray of light. What about the converse: Could light be made to electrify or magnetize objects? On a bright day, he shone a beam of sunlight down the length of a helical coil. No effect. He put a bar of unmagnetized steel inside the coils, but there was still no effect, even when he tried rotating the steel bar. This was one of hundreds of failed attempts at finding connections between one type of force and another—in some others he tried to connect electricity and magnetism with gravity. Scientists are still looking for that link, as part of their search for a single theory that will unite the four forces known today—the electromagnetic force, the weak nuclear force, the strong nuclear force, and gravity. (The first two of these have been unified in something called the *electroweak force*.)

The Royal Institution's Friday Evening Discourses had by now become an institution in their own right. The lecture on April 3, 1846, turned out to be a historic occasion, although none of the audience recognized it as such and the whole thing happened by chance in rather bizarre fashion. Charles Wheatstone was to have been the latest in a long line of distinguished speakers, but he panicked and ran away just as he was due to make his entrance. Although amply confident in his professional dealings as scientist, inventor, and businessman, Wheatstone was notoriously shy of speaking in public, and Faraday had taken a gamble when engaging him to talk about his latest invention, the electromagnetic chronoscope—a device for measuring small time intervals, like the duration of a spark. The gamble had failed, and Faraday was left with the choice of sending disappointed customers home or giving the talk himself. He chose to talk, but he ran out of things to say on the advertised topic well before the allotted hour was up.

Caught off-guard, he did what he had never done before and gave the audience a glimpse into his private meditations on matter, lines of force, and light. In doing so, he drew an extraordinarily prescient outline of the electromagnetic theory of light, as it would be developed over the next sixty years. In his vision, shared by nobody else at the time, the universe was crisscrossed by lines of force—electric, magnetic, and possibly other kinds. The points where these lines met were the points at which we *perceive* matter to exist; his "atoms" were merely the centers of forces that extended through all space. When disturbed, the lines of force vibrated laterally and sent waves of energy along their lengths, like waves along a rope, at a rapid but finite speed. Light, he suggested, was probably one manifestation of these vibrations. He was emphatic that the vibrations were vibrations *of the lines of force themselves*, not of the supposed luminiferous aether—the imponderable medium that was thought necessary to transmit light waves. Faraday was doubtful that such an aether existed; he commented that it would have to be "destitute of gravitation and infinite in elasticity."

As if all this were not enough, he suggested that gravity, too, might be propagated by lines of force:

> The propagation of light and therefore of all radiant action occupies time; and a vibration of the line of force should account for the phenomena of radiation, so it is necessary that such vibration should occupy time also. I am not aware whether there are any data

> by which it has been, or could be ascertained, whether such a power as gravitation acts without occupying time or whether lines of force being already in existence, such a lateral disturbance of them at one end . . . would require time, or must of necessity be felt at the other end.[5]

To his contemporaries, the idea that gravity could act through lines of force and might be somehow connected to electricity and magnetism seemed bizarre. In their eyes, gravitation was, by the law of Newton, a rectilinear force that acted instantaneously at a distance; electricity and magnetism were fluids; and light was a vibration of an imponderable substance. All this could be explained in elegant mathematics, yet here was a mathematical illiterate putting forward ideas that, if taken seriously, would threaten to overturn the established laws of the physical world. From today's perspective, it is clear that this was a historic moment. Faraday, the bold theorist, was making an advance announcement of a scientific transformation that has given us not only electromagnetic theory but also special relativity, radio, television, and much more besides.

But that was not how things seemed at the time. Faraday may have regretted that he had let these private thoughts escape, so opening himself to ridicule, but the damage had been done. To limit it, he recapitulated the talk in an article for *Philosophical Magazine* called "Thoughts on Ray-vibrations," which closed with the passage:

> I think it likely that I have made many mistakes in the preceding pages, for even to myself my ideas on this point appear only as a shadow of a speculation, or as one of those impressions on the mind which are allowable for a time as guides to thought and research. He who labors in experimental inquiries knows how numerous these are, and how often their apparent fitness and beauty vanish before the progress and development of real natural truth.[6]

The general scientific opinion was that Faraday had ventured far out of his depth. His peers recognized that he was a superb experimentalist but felt that, having no mathematics, he was simply not equipped for any kind of theorizing. His latest outpouring seemed amply to confirm this view. Even Faraday's supporters were embarrassed. His first biographer, Henry Bence Jones, dismissed the "Thoughts on Ray-vibrations" paper in half a line; his third, John Hall Gladstone, didn't mention it at all; and his second, John Tyndall, who succeeded him at the Royal Institution, described it as "one of the most singular speculations that ever emanated from a scientific man."[7] Yet, speculative as they were, Faraday's ideas were essentially correct, as Maxwell and his followers would show.

Whatever they thought of Faraday's ability as a theorist, scientists everywhere were stunned by his discovery of diamagnetism. Why did substances behave in this unexpected way? Hans Christian Oersted, now a grand old man of science, suggested reverse polarity: a magnetic pole induces a pole of the opposite kind in a paramagnetic substance like iron but one of the same kind in a diamagnetic substance like bismuth. Edmond Becquerel, whose father had worked with Ampère, came up with an ingenious theory analogous to Archimedes's principle in hydrostatics: All substances had magnetic power, but the more powerful ones (the paramagnetics) always tended to displace the weaker (the diamagnetics), just as a dense liquid displaces a less dense body like an air bubble. A neat answer, but to explain why diamagnetics were repelled by magnetic poles even in a vacuum Becquerel had to invoke an all-pervading aether that had its own degree of magnetism: less than the paramagnetics but more than the diamagnetics.

In the course of his own investigations, Faraday, as was his way, carefully examined everyone else's theories and experiments. Oersted's reverse-polarity hypothesis was first on his list. He sprinkled filings of bismuth on paper over a magnet, reasoning that if Oersted was right, they would arrange themselves in the familiar pattern that iron filings did—each little filing would be the wrong way around, as it were, but the pattern would look the same. In other words, if the bismuth filings showed the same pattern as iron filings, they would reveal the presence of poles. However, Faraday's bismuth filings showed no signs of lining up. Oersted seemed to be wrong; there was no sign of reverse polarity.

Then he heard about a brilliant experiment by the German physicist Wilhelm Weber, at that time a professor in Leipzig, that made him think again. It was a clever variation on Faraday's own iron-ring experiment, with a primary coil and a secondary one, but without the ring. A helical primary coil with a soft iron bar inside it was connected to a battery, forming a strong electromagnet. The secondary coil was another helix, coaxial with the first and a short distance away. It was wound on a wooden tube so that a bar, either of iron or of bismuth, could be pushed into the coil and pulled out again. The secondary circuit was completed by connecting the ends of the coil to a galvanometer, which would indicate the direction of any current. When Weber pushed the iron bar into the secondary coil, the galvanometer needle twitched briefly to the right, and when he pulled it out, the needle did the same but this time to the left. All as expected. But when he substituted the bismuth for the iron, the needle twitched *the other way*—a clear demonstration of reverse polarity.

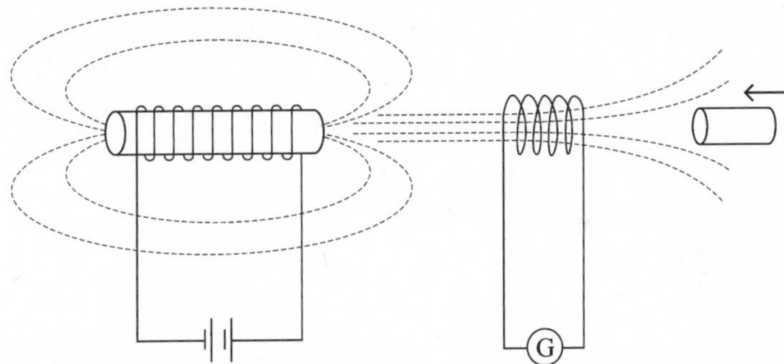

Fig. 6.2. Schematic layout of Weber's experiment. The wooden tube on the right held either a bar of iron or a bar of bismuth. (Used with permission from Lee Bartrop.)

Faraday thought of a way to explain Weber's experiment. He had long believed that magnetism acted along lines of force. Was it not likely that some substances offered easier passage to the lines than others? He had shown that each substance had its own inductive capacity for static electricity; should it not similarly have its own capacity for conducting magnetic lines of force? Everything began to fall into place. Paramagnets conducted lines of force much better than the surrounding air did, so the lines converged on them. Diamagnets, on the other hand, conducted the lines more poorly than air, so the lines diverged from them.

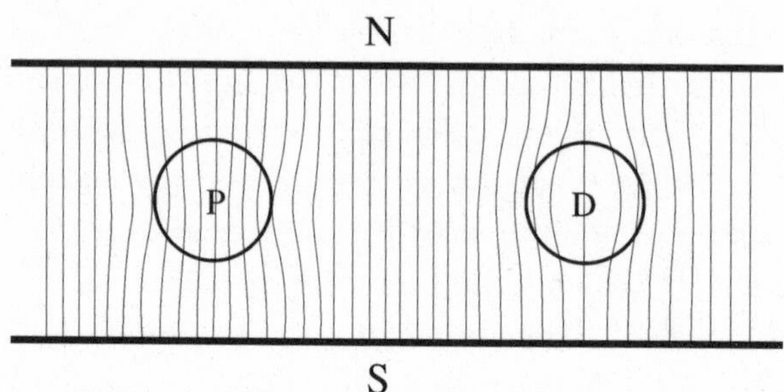

Fig. 6.3. Magnetic lines of force converge through a paramagnetic substance *(left)* **and diverge through a diamagnetic one** *(right)*. **(Used with permission from Lee Bartrop.)**

Consequently, in air, paramagnets tended to move to regions where the lines were denser and the magnetic force stronger, and diamagnets to regions where the lines were more sparse and the magnetic force weaker. In general, the magnetic behavior of a material object would depend on the medium that surrounded it. If it conducted lines of force better than the surrounding medium, it would act like a paramagnet; and if it conducted lines of force less well than the surrounding medium, it would act like a diamagnet. So simple. There was no poles at all, in the sense of centers of force. There was no attraction or repulsion, and certainly no action at a distance; all that happened was that bodies reacted to the patterns of lines of force in their own localities—in other words, to the field.

Weber's experimental result was simply a consequence of the grand design. When the iron was pushed into the secondary coil, the lines of force converged on it, thus cutting the wires in the coil and inducing a current. When, instead, the bismuth was pushed in, the lines of force diverged from it, again cutting the coil but this time inducing a current in the opposite direction.

Diamagnetism had to wait for atomic theory and quantum

mechanics for a full explanation. Broadly speaking, all substances are diamagnetic, but some are also paramagnetic and their paramagnetism swamps the feeble diamagnetism. Iron, cobalt, and nickel possess a further quality, now called *ferromagnetism*, which swamps both the other forms of magnetism and enables the metal to form a permanent magnet. Although their theories were deficient, Oersted, Becquerel, and Weber all emerge with credit. Faraday's theory turned out to be remarkably close to Becquerel's hydrostatic analogy, and he enjoyed the collegial rivalry with Weber, a fellow seeker of truth. When he had finished his paper, Faraday wrote to Oersted, promising to send him a copy and making a delightful observation on the whole episode:

> Is it not wonderful that views differ at first? Time will gradually sift and shape them. And I believe that we have little idea at present of the importance they may have ten or twenty years hence.[8]

Faraday had so far failed to detect magnetic properties in gases, but in 1847 he had news from Italy: Professor Michele Bancalari of Genoa University had shown that magnetism affected the behavior of flames. Faraday, as always, checked the result for himself. He expressed surprise that he had "failed to observe the effect years ago,"[9] and, within a few weeks, he had experimented with many gases and shown not only that they all possessed magnetic properties but also that they differed considerably, one from the other—most were diamagnetic, but oxygen was paramagnetic.

Part of Faraday's investigation was to study the magnetic properties of empty space, using the best vacuum pumps available. Two questions arose. First: Why did a vacuum conduct the lines of force better than diamagnetic substances did? He couldn't answer, and neither could anyone else before the development of quantum mechanics in the twentieth century. Second: How did a vacuum conduct lines of force? His answer was enigmatic:

> Mere space cannot act as matter acts, even though the utmost latitude be allowed to the hypothesis of an aether.[10]

In his "Ray-vibrations" talk, he had dismissed the aether outright. Had the hard line softened? He clarified his views a little in subsequent papers: The aether was probably an unnecessary contrivance, but if it did exist, it would serve to transmit magnetic lines of force as well as light.

For years, Faraday had been thinking heresies but, aside from his off-guard outburst in the impromptu "Ray-vibrations" lecture, he had largely kept them to himself. His reasons were clear. Even though he was recognized as a great experimental scientist—the Royal Society had awarded him three medals, including its top one, the Copley Medal—Faraday knew that his lack of mathematics was a bar to similar recognition as a theorist. The mathematically based Newtonian tradition, with its theories of action at a distance along straight lines between electric charges or magnetic poles, still held sway, and almost nobody thought that the beautiful mathematical theories of Coulomb and Ampère might be wrong. Even Faraday's brief and guarded references to the possible existence of lines of force had generally been greeted with derision. The results that came from his laboratory were, in truth, becoming harder and harder for the mathematicians to explain in Newtonian terms, but Weber had taken on Ampère's mantle, and he and others were producing ever more ingenious mathematical models to account for everything by action at a distance. By comparison, Faraday's theories seemed like a child's fantasy. The prevailing view was cogently expressed by the British Astronomer Royal, Sir George Biddell Airy:

> I can hardly imagine anyone who practically and numerically knows the agreement [between calculations based on action at a distance and experimental results] to hesitate an instant between this simple and precise action on the one hand and anything so vague and varying as lines of force on the other.[11]

But to Faraday, experimental results were the only truth. He had always been prepared to abandon any line of inquiry the moment his results showed it to be false. Yet his idea of lines of force had withstood every test. In the course of hundreds of experiments over the years, culminating with those on diamagnetism, he had come to believe that lines of force were not simply an indication of the presence and pattern of electric and magnetic forces and of the accompanying strains in matter, they were the *vehicle* that conveyed the forces and were physically present in space. Though nothing could be proved beyond doubt, the evidence from experiment after experiment overwhelmingly favored them over the rival hypothesis of action at a distance. All his results suggested that *nothing* happened at a distance, that all forces and all induction acted through some kind of strain in the intervening medium, and that the strain acted along paths that were rarely straight. After his success in discovering diamagnetism, Faraday became bolder and began to mount an open challenge to orthodox views. The term *magnetic field* now cropped up regularly in his writings and he tried to convey his ideas in one of the reports in his *Experimental Researches in Electricity*:

> From my earliest experiments on the relation of electricity and magnetism, I have had to think and speak of lines of magnetic force as representations of magnetic power—not merely in the points of quality and direction but also in quantity.[12]

This is a little disingenuous—his thinking and speaking on this controversial topic had been largely private. But he goes on:

> Important to the definition of these lines is that they represent a determinate and unchanging amount of force. Though, therefore, their forms, as they exist between two or more centers or sources of power, may vary greatly, and also the space through which they may be traced, yet the sum of power contained in any one section of a given portion of the lines is exactly equal to the sum of power in any other section of the same lines, however altered in form or however convergent or divergent they may be at the second place.[13]

Faraday, the nonmathematician, was here describing precisely a property of the magnetic field that was later given mathematical form in one of the four famous "Maxwell's equations."

In further experiments he quantified his result of 1831 that a current was induced in a wire when it "cut" magnetic lines of force, and he summed up the findings in a deceptively profound statement. The "quantity of electricity thrown into a current" was "directly as the amount of curves intersected."[14] This statement was true whether the curves were dense or sparse, converging or diverging, and neither the shape of the wire nor its mode of motion made any difference, except that the *direction* of the current depended on what became known as the right-hand rule.[15] It was the original statement of one of the most fundamental laws of electromagnetism—now called simply *Faraday's law of induction.*

One of Faraday's heresies was to question the existence of magnetic poles. He had three good reasons. Two of these were acknowledged by everyone, though only Faraday seems to have pursued their consequences. First, why should arbitrary points at the end of magnets act as centers of force? Second, nobody had ever found a single pole; they always occurred in north–south pairs. If you sliced up a magnet, every sliver, no matter how thin, would have a north face and a south face, and this suggested that the power of the original magnet was distributed along its length rather than concentrated at the ends.

The third reason was Faraday's own. Unlike lines of electric force, which ran between oppositely charged objects, lines of magnetic force didn't start or stop anywhere—they were continuous loops. His floating-needle experiment of 1831 had shown that each loop of magnetic force around a current-carrying coil ran all the way through the coil. If an iron bar magnet of the same size and strength were substituted, the magnetic effect would be exactly the same—the pattern of loops of force would be unchanged. It seemed that the loops of force must run along the length of the iron magnet just as they did through the coil. The ends of the bar magnet were simply the surfaces where the looping lines of force entered and left; there were no poles.[16]

Faraday chose not to offer any explanation of the source of the lines of force inside a permanent magnet. One may wonder why he didn't accept Ampère's and Fresnel's hypothesis of tiny currents that circulated around the iron particles in a magnet, as this would have fit in so well—each little loop of current would have a line of force running through it. Perhaps he didn't want to sully the concept of lines of force, perfect in itself and supported by hard-won experimental evidence, by adding on something as speculative as Ampère's and Fresnel's currents. Moreover, he was a chemist and perhaps something about the idea of electric currents circulating around particles of matter didn't seem right to him.

Along with magnetic lines of force went another notion. Faraday wrote, "Again and again the idea of an *electrotonic state* has been forced on my mind."[17] He could explain all the known interactions of magnetic fields with electrical circuits very well in terms of lines of force but couldn't shake off the idea that some kind of state of strain or tension was involved in the process. Time and again, he observed something first shown in his iron-ring experiment of 1831. When a magnetic field was suddenly removed from the neighborhood of a wire loop, a brief current flowed in the loop. This could be explained by lines of force collapsing and "cutting" the wire, but it also seemed to Faraday that a state of strain existed in and around the wire loop, and that the release of this strain took the form of a current. Similarly, the brief current that accompanied the discharge of a storage device like a Leyden jar seemed to come from the release of a strain in the insulating material between the charged metal surfaces—the electrical version of the electrotonic state.

In 1855, Faraday completed the third and last volume of his *Experimental Researches in Electricity*.[18] These volumes were a faithful account of hundreds of investigations, both successful and unsuccessful; of his many bold hypotheses, most discarded or modified after rigorous testing; and of his chains of reasoning—all pervaded by a spirit of relentless inquiry. The work had been accomplished despite an increasingly failing memory—he once found when looking back at his notes that he had repeated a long series of experiments done only six months earlier—and a dwindling stock of the mental energy that had been so abundant in his youth. He had coped by placing bounds on his life, taking frequent breaks from laboratory work and turning down almost all invitations to dinners and other functions, yet he managed to take up social causes with all his old passion, for example, campaigning to rid the River Thames of pollution.

He also did his best to expose charlatans who promoted the mystic nonsense of table-turning that gripped fashionable London in the 1850s—when several people placed their hands on a table it would move, as though propelled by ghostly powers. Faraday personally investigated three séances where the participants sat in the dark with their hands palms down on the table top and mystic forces supposedly turned the table. When the experiments were tried again after Faraday had placed hidden rollers to detect pushing, the table, unsurprisingly, remained immovable. Equally unsurprisingly, the fraudsters conducting the séances shrugged off this outcome, claiming that one could hardly expect spirits to perform while under close surveillance. In July 1853, Faraday wrote a long letter to the *Athenaeum* magazine debunking the fad, causing the spiritists and their disturbingly large body of sympathizers to rail bitterly against him. He wrote to a friend:

> I have not been at work except in turning the tables upon the table turners nor should I have done that, but that so many inquiries poured in upon me, that I thought it better to stop the inpouring flood by letting all know at once what my thoughts and views were. What a weak, credulous, unbelieving, superstitious, bold, frightened, what a ridiculous world ours is, as far as concerns the minds of men. How full of inconsistencies, contradictions and absurdities it is.[19]

Despite the efforts of Faraday and others, the superstition persisted. Nine years later, he turned down an invitation to a séance, saying, "I will leave the spirits to find out for themselves how they can move my attention. I am tired of them."[20]

Faraday had made great discoveries, but in the 1850s it was by no means apparent where they would lead. He had demonstrated the principles of the electric motor, the electric generator, and also, by his iron ring, the transformer—a device now indispensable in electrical power systems—but they had so far not yielded much. Inventors had produced electromagnetic machines, but they were largely curiosities, of little practical use. What eventually opened up the field was, curiously, the development in the 1870s by Heinrich Geissler and others of effective vacuum pumps. This made possible the filament light bulb, which, in turn, led to investment in power-distribution systems requiring efficient generators. It then became practicable to develop electric motors for all manner of purposes, fed from the grid. And in the late 1800s, Nikola Tesla demonstrated the advantages of high-voltage, alternating-current power-distribution systems, which required transformers.

Electric lighting may have been some way off, but another application of electricity was the wonder of the age. Pioneered by Charles Wheatstone and his partner William Fothergill Cooke in Britain and by Samuel Morse in America, the telegraph was making spectacular progress. By the late 1840s, many cities were already connected by landlines, but telegraphers couldn't get lines to work reliably under water. Among his many other activities, Faraday played a part in making undersea telegraphy possible. The problem had been to find a satisfactory material to use as an insulator between the metal cable core and its outer sheath, which was in contact with the seawater. Telegraphers had tried everything they could think of—cotton, rubber, rope soaked in boiled tar—but the insulation always broke down. Then, in 1848, somebody sent a sample of a Malayan tree gum called gutta percha to Faraday, who tested it and found it to be not only an excellent insulator but also water-resistant, flexible, resilient, and easily molded to shape when warm. Such was Faraday's reputation that his recommendation quickly led to a huge demand for gutta percha. The material met every demand placed on it, and cables soon ran across the English Channel and the Irish Sea.

The next challenge was both obvious and daunting—to lay a cable under the Atlantic Ocean. The task presented many problems, one of which was that the signal pulses became blurred when sent along undersea cables; signals had to be sent slowly so that each pulse could be distinguished from the next. The longer the cable, the worse the problem, so before investing a huge sum in its cable project, the Atlantic Telegraph Company needed to know whether it would be able to send signals at a fast-enough rate to be profitable.

At first, telegraphers had no idea why the blurring happened, but Faraday soon supplied the answer. A cable, with its copper core surrounded by insulating material and an outer sheath, was rather like a hugely extended Leyden jar. He had found in the laboratory that all electrical induction takes time to act through the insulating medium. Hence a device like a Leyden jar takes time to charge and to discharge; and this was exactly what was happening in the cable. When the telegraph operator pressed his key at the sending end, the current at the receiving end grew only gradually and didn't reach full value until the cable was fully charged. And when the key was released, the current at the receiver similarly took time to fall to zero. So what should have been a sharp pulse turned into a smeared-out blob.

But Faraday couldn't quantify the charging time, and the Atlantic Telegraph Company turned to William Thomson, the young Scot who had earlier worked out an equation for Faraday's lines of electric force. He had then used the analogy of heat flow and now did so again, reasoning that electrical induction would diffuse through the insulating material like heat through a metal bar. This way, he worked out an equation for undersea cables and gave the company the bad news that the charging time increased proportionally with the *square* of the distance—even using an expensive, low-resistance cable, it would take far longer to send a message across the Atlantic than across, say, the Irish Sea. They went ahead anyway and, after many setbacks, laid a cable in 1858, only for it to fail shortly after transmitting messages between Queen Victoria and President Buchanan at a rate of two seconds per character. Later attempts were more successful: the company eventually made a profit for its staunch shareholders, and Thomson earned a knighthood.

The mercurial Thomson rarely worked on any topic for more than a few weeks before his mind was captured by a new idea, often in a completely different area of science. His early study of Faraday's electric lines of force had been a typical inspired burst. From time to time he returned to electricity; he gave mathematical expression to Faraday's idea that each material had its own specific inductive capacity for electricity and another for magnetism, and he derived a formula for the total energy of a magnetic system. But, as we'll soon see, his greatest contribution was to give some good advice to another young Scot who wanted to study electricity.

Faraday was a lone worker. He had no close colleagues after Davy and, although he had been Davy's protégé, he never took on a pupil of his own. In part-explanation, Faraday wrote:

> I have looked long and often for a genius for our laboratory, but have never found one. But I have seen many who would, I think, if they had submitted themselves to a sound self-applied discipline of mind, have become successful experimental philosophers.[21]

He had nevertheless turned down an ardent request from Lord Byron's daughter Ada Lovelace, who wanted to join him and become, in her words, a bride to science. Perhaps he was not prepared for Lovelace's particular brand of dedication. Or perhaps his intensely personal style of working made it impossible for him to take on the role of mentor.

Whatever the reasons, it seemed that Faraday's radical ideas about physical lines of force in space might wither on the vine. No one but Thomson had shown any interest, and he regarded lines of electric force principally as an interesting aspect of the mathematics. Nevertheless, Thomson made a vital contribution: When a young friend asked for advice on how best to start studying electricity, he told him to be sure to read Faraday's *Experimental Researches in Electricity*.

Early in 1857, Faraday received in his post a copy of a paper with the title "On Faraday's Lines of Force." The author was James Clerk Maxwell, a young professor of natural philosophy at Marischal College, Aberdeen, who had written the paper while still a student at Cambridge. It began:

> The present state of electrical science seems peculiarly unfavorable to speculation.[22]

Faraday must have feared for what was to come, but this was Maxwell's way of introducing his use of an analogy from another branch of physics. The analogy he presented for lines of force was the flow of an incompressible fluid. The streamlines of flow represented lines of force, either electric or magnetic, while the speed and direction of fluid flow at any point represented the *density* and direction of the lines of force there. He was able to extend the analogy to cover all static electrical and magnetic effects, including the magnetic force between two current-carrying circuits. The mathematics of fluid flow were indisputable, and he set them out in words with a few equations by way of summary. In the final sentence of part 1 of the paper, Maxwell signaled his intention to take Faraday's guidance on electromagnetism:

> I hope to discover a method of forming a mechanical conception of this electro-tonic state adapted to general reasoning.

In part 2 he did, indeed, find a way to describe the electrotonic state mathematically. We can imagine Faraday's joy. Here, at last, was somebody prepared to take on his ideas and work with them. He replied:

> I received your paper, and thank you very much for it. I do not venture to thank you for what you have said about "Lines of Force" because I know you have done it for the interests of philosophical truth; but you must suppose it is work grateful to me, and gives me much encouragement to think on. I was at first almost frightened when I saw the mathematical force made to bear on the subject, and then wondered to see that the subject stood it so well.[23]

Perhaps emboldened by Maxwell's support, Faraday published a paper formally proposing *gravitational* lines of force, extending an idea first mentioned in his "Ray-vibrations" talk. He knew that the reaction of the great body of his fellow scientists would be disbelief, and he may have felt some anxiety when he asked Maxwell for an opinion. He needn't have worried. Maxwell sent a long and thoughtful reply, concluding that the idea was sound and that the gravitational lines of force could "weave a web across the sky" and "guide the stars in their courses."[24] To this, Faraday responded:

Your letter is the first intercommunication on the subject with one of your mode and habit of thinking. It will do me much good, and I shall read and meditate it again and again. . . . I hang on to your words because they are to me weighty and . . . give me great comfort.[25]

Faraday had found his successor.

CHAPTER SEVEN

FARADAY'S LAST YEARS

1857–1867

Faraday's achievements during his last decade are not of the kind that make the history books. They are nevertheless remarkable, accomplished against the background of both ever-diminishing powers of memory and spells of nervous exhaustion that eventually made even the simplest mental task wearisome.

Much of the work was inspired by a strong sense of public duty, nowhere better shown than by his service to Trinity House. He had given up trade and a profitable career as a consulting chemist to pursue the life of a natural philosopher, whose thoughts were elevated to the highest peaks of abstraction, far removed from the mundane daily routine. In his later years, with his unruly white hair, he even looked the part of the unworldly, absentminded professor. Yet, with his work for Trinity House, he became a practical-minded Victorian businessman, alert to budgetary matters and hard on anyone who proposed unnecessary or wasteful innovations to lighthouse operations.

During the late 1850s, he wrote more than twenty reports for Trinity House. They covered all aspects of lighthouse operation, but the main concerns were the brightness and reliability of the lights, and one of Faraday's tasks was to oversee the experimental introduction of electric light as an alternative to oil or gas-fueled lamps. After many preliminary tests, a full operational system, though still an experimental one, was installed at South Foreland, near Dover, in 1858. A magneto-electric generator driven by a steam engine supplied current to a carbon-arc lamp, and the first electric light shone across the English Channel. Several similar electric systems followed at other lighthouses, but their high cost and less-than-perfect reliability caused the electrification program to be put off until the 1920s, when filament light bulbs and central generation of electricity at last made it a practical proposition.

Faraday made frequent inspection visits to lighthouses and went out in all kinds of weather on the Trinity House vessel to test the visibility of the lights from the sea. Such work would have tested the mettle of a man half his age, as is evident from a report he made in 1861 of a routine visit to South Foreland. He was seventy years old.

> I went to Dover last Monday; was caught up in a snowstorm . . . could not go to the lighthouse that night; and finding next day that the roads on the downs were snowed up returned to London. On Friday I again went to Dover . . . hoping to find the roads clear of snow; they were still blocked up towards the lighthouse, but by climbing over hedges, walls and fields, I succeeded in getting there and making the necessary enquiries and observations.[1]

It is difficult for us today to appreciate the immense importance then attached to saving lives (and cargoes) at sea. As late as 1912, the Nobel Prize in physics went to Niels Gustav Dalen for inventing a way of feeding gas automatically to lighthouses and buoys. His achievement, in the committee's judgment, had surpassed those of rival nominees Albert Einstein, Max Planck, Hendrik Antoon Lorentz, Ernst Mach, and Oliver Heaviside. Faraday's lighthouse work was a service to his fellow men, wholeheartedly given and well appreciated. And, rather appropriately, one of Faraday's comments on lighthouses illuminates for us how religious faith inspired his whole approach to scientific work. In a report, he wrote:

> There is no human arrangement that requires more regularity and certainty of operation than a lighthouse. It is trusted by the Mariner as if it were a law of nature, and as the Sun sets so he expects that, with the same certainty, the lights will appear.[2]

In the laboratory he sought to reveal more of God's laws of nature, and in his work for Trinity House he treated these laws as a model to be emulated to the highest degree possible.

Faraday was the man the government turned to for practical scientific advice on any topic. During the Crimean War, the War Office had consulted him on what would by today's rules have been a highly classified matter. They asked him about the likely formation and movement of clouds of poison gas, a weapon then being considered to help defeat the Russians. In replying, he drew on his memories of Mount Vesuvius from the Grand Tour of half a century earlier, when a treacherously changing wind had blown noxious fumes from the crater in his direction, almost choking him. Evidently, poison gas was a double-edged weapon; the War Office decided not to use it.[3] This episode suggests that Faraday could remember events of fifty years ago more clearly than those of last week.

He advised on how best to preserve paintings in the National Gallery in London and artifacts such as the Elgin Marbles at the British Museum. He also did his best to encourage schools to improve the woeful education in science that most of them provided. He was appalled at the failure of otherwise well-educated people to understand even the simplest scientific principles. On one occasion he complained to the public-school commissioners:

> They come to me and they talk to me about things that belong to natural science; about mesmerism, table turning, flying through the air, about the laws of gravity; they come to me to ask questions, and they insist against me, who think I know a little of these laws, that I am wrong and they are right, in a manner that shows how little the ordinary course of education can teach these minds. . . . They are ignorant of their ignorance . . . and I say again there must be something wrong in the system of education which leaves minds, the highest taught, in such a state.[4]

The authorities were, indeed, ignorant of their ignorance. Faraday might as well have been talking to the wall: Latin and Greek, with a smattering of Euclid, remained the staple diet in English public-school education for some time.

To the end he continued to seek scientific truth. Over the years, he became more and more convinced of the unity of all nature's forces and an imperative was to search for evidence that the force of gravity was linked to those of electricity and magnetism. His experiments on gravity were heroic failures. Back in 1849 he had rigged up a helix of copper wire 350 feet long, with its axis vertical, and dropped blocks of various substances through it—from the high ceiling of the Royal Institution's lecture theater to a cushion on the floor. He reasoned that a material body held stationary amid gravitational lines of force might be in a state of strain akin to the electrotonic state of a metal wire amid magnetic lines of force. If so, then perhaps the strain would be relieved when the body was allowed to fall freely, and this might—by analogy with electromagnetic induction—result in a current in the coil, which could be detected with a galvanometer. He tried dropping blocks of iron, copper, bismuth, and other materials. No current, but the results didn't shake his "strong feeling of the existence of a relation between gravity and electricity," and in 1859 he returned to the task.[5]

This time he tried raising and lowering huge, electrically charged blocks of lead through the longest vertical distance he could find, to see if the charge varied. After considering the tower at the Houses of Parliament, he settled on the 165-foot shot tower[6] near Waterloo Bridge for his experiment. There was no significant variation in the charge, but Faraday wrote up the results and submitted them to the Royal Society. On this topic, Faraday thought, even a negative result was important enough to publish. The Society's secretary, George Stokes, did not agree. Like most of his colleagues, he never took Faraday's ideas on the unity of forces seriously, and he believed (correctly) that he was saving the revered old man from ridicule by advising him to withdraw the paper. Faraday acquiesced.

His very last experiment was performed in March 1862, when he investigated the effect of a magnet on the light spectra of incandescent substances—his mind was still testing the boundaries of what was physically possible. He lit a gas flame between the poles of a magnet and looked for optical effects.

> The colorless Gas flame ascended between the poles of the Magnet and the salts of Sodium, Lithium, etc. were used to give color. A Nicol's polarizer was placed just before the intense magnetic field and an analyzer at the other extreme of the apparatus. Then the E Magnet was made and unmade but not the slightest trace of any effect on or change of the lines in the spectrum was observed in any position of the polarizer or analyzer.[7]

Another failure. But, to us, Faraday's genius and vision shine through. Here, as in his quest to unify all known forces, he was sowing a seed for a harvest to be gathered by future scientists. In 1897, the Dutch physicist Peter Zeeman repeated the experiment, using a stronger magnetic field and a more refined apparatus, and he found the very effect that Faraday had been looking for. The Zeeman effect, as it is known today—the splitting of the light spectrum into several components in the presence of a magnetic field—makes possible such techniques as magnetic resonance imaging. We can only wonder at the man who, even with fading mental powers, was able to envisage this effect of magnetism on light.

As Faraday's health and mental faculties declined, he began to relinquish his various responsibilities at the Royal Institution, finally handing over the directorship to John Tyndall in 1865. The consequent loss of income, and of his flat, would have been a worry, but in 1858 Prince Albert, a great admirer, had asked the queen to put a house at Hampton Court at his disposal. Faraday had refused at first, fearing the high cost of repairs, but the queen said she would pay. He and Sarah moved in, and the new house became his last home.

He maintained his resolution not to be distracted from his scientific investigations by commercial work or by accepting high office. The Royal Society asked him to accept their presidency in 1857. Faraday declined. His health was not up to it, and the bitter taste from his election to F. R. S. back in 1824 had not entirely faded; when asked for a list of the many scientific honors he had received over the years, he commented:

> One title, namely that of F. R. S. was sought and paid for; all the rest are spontaneous offerings of kindness and goodwill from the bodies named.[8]

A rumor once got around that Faraday had been knighted. When somebody wrote asking about it, Faraday replied:

> I am happy that I am not a Sir, and do not intend (if it depends upon me) to become one.[9]

There is little doubt that, had he wished, Faraday could have become a knight and probably, later, a baron, as William Thomson did. But he was more receptive to honors from overseas. Following the precedent of Davy's award of the Napoleon Prize,[10] he accepted from Napoleon III the title of Commandeur de la Légion d'Honneur. This was by no means the only foreign title he held. He was a Chevalier of the Prussian Order of Merit and, by order of the Royal House of Savoy, a Knight Commander of the Order of St. Maurice and St. Lazarus. One may wonder why plain Michael Faraday, who held to the simple life, scorned pomp, and spurned civic honors in his own country, was happy to accept these grand titles from elite bodies overseas. He gave a partial explanation, saying "By the Prussian knighthood I do feel honored; in the other I should not."[11] Perhaps remoteness played a part, too: he knew he would never be called upon to wear the lavish robes or take part in the elaborate ceremonies.

Until his final retirement, Faraday traveled from Hampton Court to the Royal Institution most days and, from 1860, had a new visitor. James Clerk Maxwell had taken up a post at King's College in the Strand and lived in Kensington, so his daily journey to work took him close to Albemarle Street. He attended some of the Friday Evening Discourses and, at Faraday's invitation, gave a memorable one himself. There is no documentary record of their having met informally, but we can be fairly certain that they did, and it is pleasing for us to picture them together, two modest and genial men whose combined endeavors changed the world.

Gradually, from 1862 onward, Faraday's health deteriorated and his mental grasp of what was going on around him crumbled; the present and the past were equally confused in his mind. In a last letter to a close friend, he wrote:

> My Dear Schönbein,
>
> Again and again, I tear up my letters, for I write nonsense. I cannot spell or write a line continuously. Whether I shall ever recover—this confusion—I do not know. I will not write anymore. My love to you.[12]

Willpower was no longer enough to keep a hold on day-to-day events, and the haze that he had so far been able to dispel now enveloped him. He spent most of his days sitting silently, and on fine evenings enjoyed watching the sunset through the window. He died peacefully in his chair in August 1867. In accordance with his wishes, his funeral was "strictly private and plain."[13] He was buried in Highgate Cemetery, and the headstone at his grave reads, simply:

Michael Faraday
Born 22 September 1791
Died 25 August 1867

True to his words, he remained plain Michael Faraday to the end.

Heartfelt tributes to Michael Faraday came from all over the world. They could fill a book. One from John Tyndall, who knew him as well as anybody, gives us an extra insight into Faraday's character:

> We have heard much of Faraday's gentleness and sweetness and tenderness. It is all true but it is very incomplete. You cannot resolve a powerful nature into these elements, and Faraday's character would have been less admirable than it was had it not embraced forces and tendencies to which silky adjectives "gentle" and "tender" would by no means apply. Underneath his sweetness and gentleness was the heat of a volcano. He was a man of excitable and fiery nature, but through high self-discipline he had converted the fire into a

> central glow and motive power of life, instead of permitting it to waste in useless passion. "He that is slow to anger" saith the sage, "is greater than the mighty and he that ruleth his own spirit than he that taketh a city." Faraday was *not* slow to anger, but he completely ruled his own spirit, and thus, though he took no cities, he captivated all hearts.[14]

Despite the universal acclamation of Faraday's scientific work, his greatest achievement had been largely ignored during his lifetime and was only beginning to surface at the time of his death. The great German physicist Hermann von Helmholtz made this tribute in 1881:

> Now that the mathematical interpretation of Faraday's conceptions regarding the nature of electric and magnetic forces has been given by Clerk Maxwell, we see how great a degree of exactness and precision was really hidden behind the words which to Faraday's contemporaries appeared either vague or obscure; and it is in the highest degree astonishing to see what a large number of general theorems, the methodical deduction of which requires the highest powers of mathematical analysis, he found by a kind of intuition, with the security of instinct, without the help of a single mathematical formula.[15]

A man of equal stature and complementary talents was needed to reveal Faraday's full greatness. That man was James Clerk Maxwell.

CHAPTER EIGHT

WHAT'S THE GO O' THAT?

1831–1850

Far away from the smoky, noisy streets of Faraday's London, the Vale of Urr nestles amid the gently rolling hills of Galloway in southwest Scotland. It was there that young James Clerk Maxwell took his first steps and spoke his first words. His father had inherited an estate that came to be called Glenlair, a tranquil and beautiful place, its fields and woods traversed by the rippling waters of the Urr. Though Maxwell spent much of his life elsewhere, he remained a country boy at heart, rooted in the land and at one with those who worked on it. Glenlair was more than a charming piece of countryside; it was his home—a source of inspiration and of solace when needed. To understand Maxwell, we need to look a little into the history of Glenlair and its people.

Things had not always been so peaceful. A couple of centuries earlier, the Galloway estate, then much larger and called Middlebie, was one of the strongholds of the fierce Maxwell clan that had ravaged the border country in bitter rivalry with the Johnstones. The Clerks of Penicuik, near Edinburgh, were, by contrast, a distinguished family with impeccable credentials. The two seemingly incompatible dynasties came together in the mid-1700s when the Clerks acquired Middlebie by marriage, on condition that whoever inherited the estate would add Maxwell to his name. As the Clerks already had their own baronetcy of Penicuik, they arranged that Penicuik would be passed to the senior heir and Middlebie to the second. So it came about that James's father was John Clerk Maxwell of Middlebie while his uncle was Sir George Clerk of Penicuik.

John's grandfather was the first Clerk Maxwell to be laird of Middlebie, but he never lived there, and when he suffered heavy losses from mining investments, most of the vast estate had to be sold, leaving a 1,500-acre residue. This passed to John's father, but he joined the British East India Company's navy and never lived at Middlebie either. When John came into the inheritance while still at school, Middlebie was, to the family, no more than a primitive outpost, and it seemed that he, too, would be an absentee landlord. He grew up in sophisticated Edinburgh society and became a lawyer but never pursued the profession with much vigor. With an adequate private income he could afford to indulge a passion for science and engineering. He enjoyed keeping abreast of the latest ideas, which might be on anything from water-treatment plants or mining technology to the mass production of teapots, and he acquired a wide circle of like-minded friends in industry, agriculture, and universities. With his closest friend, John Cay, he tried to invent and market various useful new devices, such as a bellows that would supply a continuous, steady blast. These plans came to nothing, but another began to form in his mind—to go to live at Middlebie and there apply his modern ideas on forestry and farming. Perhaps that, too, would have gone the way of other "best-laid schemes o' mice an' men," had his long friendship with John Cay's sister Frances not blossomed into love.[1] Frances was a resolute woman who supplied the get-up-and-go that he had so far lacked. She agreed to marry him, and they decided to make their life together in Galloway.[2]

It was a herculean project. The estate had been long neglected, and much work was needed, such as clearing scrub and stones, before plowing and planting could begin. There was no suitable dwelling at Middlebie, but that didn't matter—to John, the prospect of designing and building his own house was irresistible. He drew up plans for a grand mansion, but ground clearing had eaten deep into the project budget and they could only afford to build one section—the rest would have to wait. The Clerk Maxwells brought new life to Middlebie, and their new house, though small, was at its heart. They called it Glenlair, and the name soon came to be applied to the whole estate.

Frances gave birth to a daughter, Elizabeth, but joy changed to anguish when the baby died. When Frances became pregnant again, they decided to go to Edinburgh for the birth, to be near relations and a hospital if needed. This time it was a boy—James Clerk Maxwell was born on June 13, 1831. There were joyful celebrations with relations in Edinburgh, but Glenlair soon drew them back. It was now a family home, a wonderfully happy one for their son, and they watched over his development with indulgent devotion.

He soon showed himself to be a remarkable child. Nothing that went on escaped his attention. All parents have to answer incessant questions, but to be interrogated by young James was an experience of a different order. Anything that moved, shone, or made a noise drew the question "What's the go o' that?" and if the answer didn't satisfy him, he'd follow up with "but what's the *particular* go of it?"[3] In a letter to her sister, Jane, in Edinburgh, Frances describes James at age three:

> He is a very happy man . . . he has great work with doors, locks, keys, etc., and "Show me how it doos" is never out of his mouth. He also investigates the hidden courses of streams and bell-wires . . . and he drags papa all over to show him the holes where the wires go through.[*]

Recalling her own visits to Glenlair, Aunt Jane used to remark fondly that it was humiliating to be asked so many questions she couldn't answer by such a young child. He quickly learned to read and found that books not only answered some of his questions but were a delight besides; Shakespeare and Milton became particular favorites. There was the Bible, too; religion was important, and family prayers were part of the daily routine. He insisted on having a go at household chores, taking a hand with baking and basket making, and every morning helping Sandy, the gardener and handyman, to fetch water by cart from the river. Most of James's time, though, was spent in the surrounding countryside, where the rocks, the woods, the river, and the creatures that lived all around were endless sources of fascination. He ran around with the estate children, learned their

Galloway speech, and acquired a local accent that remained with him all his life.

There were no formal trappings in the Clerk Maxwell household, and James had a much-closer relationship with his parents than was usual among the gentry. His mother was his tutor, and his father often took him along on local business, chatting as though to a younger brother. John Clerk Maxwell was now well known and well liked in the Happy Valley, as the Vale of Urr was known to its residents, and an observer has left us a picture of him as he must have appeared to all who knew him:

> While . . . unostentatious and plain in all his ways, he was essentially liberal and generous. No one could look in his broad face beaming with kindliness and think otherwise. . . . By his ever-wakeful consideration he breathed an atmosphere of warm comfort and quiet contentment on all (including the dumb animals) within his sphere.

James came to understand his father's grand project, which was to go on improving the estate and the lives of all who lived there—and later took it on as his own. The Clerk Maxwells took a full part in the Happy Valley social life; there were dances and fairs, and, in the summer, picnics and archery. Life at home was busy yet harmonious and relaxed, and it hummed with jokes and gently irreverent banter—no person or institution was above some amiable debunking. The spirit of these times stayed with Maxwell all his life—he always loved a joke, and more straitlaced colleagues sometimes failed to see that a remark was meant in jest.

The idyll came to an end when James was eight years old. Frances was diagnosed with stomach cancer and, after enduring an excruciating operation without anesthesia, she died at the age of forty-seven. The family had lost its hub. Father and son were desolate, but grief drew them still closer together. Bound by the strongest of ties, the first thoughts of each were always for the other, even when they were miles apart in later years.

Maxwell's parents had originally planned for him to be home schooled until he was thirteen, when he would go straight to university, but his father was too busy with the estate and other local business to take on a tutorial role. There were no suitable schools within daily traveling distance, and John couldn't bear the thought of sending his son—his closest companion—away. James needed a new tutor, and John decided to engage a sixteen-year-old boy from the neighborhood. It was a disastrous choice. The tutor used the methods by which he had himself been taught—rote learning accompanied by physical chastisement—and it was agony for young James. Not wanting to disappoint his father, he endured the ear pulling and cuffs about the head without complaint, but nothing could induce him to learn by mechanical recitation. After a year of torment, he rebelled by rowing himself in a tub to the middle of a duck pond and refusing to come in. Aunt Jane happened to visit at about this time and was quick to realize what had been going on. Action followed promptly: the tutor was sent away and James was booked in to start at the Edinburgh Academy, one of the best schools in Scotland. John's sister, Isabella, lived only a short distance from the school, as did Jane herself, so James could stay with one or other of his aunts in term time and John could visit whenever time could be spared for the two-day coach journey from Glenlair.

James joined a class of sixty boys who had already spent more than a year at the school and developed their own pack culture. Any new boy was in for a hard time, and when this one arrived wearing a bizarre tunic and clumpy, square-toed shoes, their hostile curiosity knew no bounds. He seemed to be some kind of peasant from a far-off land, and he even spoke with a curious accent. They baited him unmercifully, and he walked home to Aunt Isabella's at the end of his first day with clothes in tatters.

The clothes were the ones he wore at Glenlair, specially designed for comfort and practicality by his father, who had done some of the tailoring and cobbling himself. Though a sagacious man in so many ways, John was prone to extraordinary lapses of judgment. Aunts Isabella and Jane promptly put this one right by seeing that the boy was properly attired for school, but the taunting went on. James bore it all with remarkable good humor, only once turning on his tormentors when goaded beyond endurance. Such bravery and composure commanded respect, and he gained a degree of acceptance. But he didn't think or behave like the other boys. He often spent his free time alone in a secluded part of the play area watching the beetles or practicing gymnastics on a tree. He drew curious diagrams and sometimes brought along homemade mechanical contraptions, but none of his fellows could make heads or tails of them. His mind whirring with impressions, questions, and partially formed ideas; he was like a steam locomotive racing away on its own while everyone else was on another track. He couldn't bring himself to join in the rote-based drudgery of the classroom and was tongue-tied when asked to perform simple oral tasks. A fish out of water, he acquired the nickname "Dafty."

Life may have been dull at school, but there was plenty of stimulation in Aunt Isabella's spirited household. James especially enjoyed the company of his elder cousin Jemima, who was a rising artist. One of their games was to make "wheels of life" for parlor entertainment: James made the rotating devices and Jemima drew sets of pictures that appeared to show a tumbling acrobat or a galloping horse when the machines were set spinning. Father and son wrote to one other regularly, and James's letters were full of whimsical banter. James addressed one letter to: Mr. John Clerk Maxwell, Postyknowswhere, Kirkpatrick Durham, Dumfries, and signed it anagrammatically as Jas Alex McMerkwell.

The rote-based drudgery in class gradually gave way to more interesting work, and young Maxwell began to take interest, rising in his second year from near the bottom of the class to nineteenth and winning the prize for scripture biography. Mathematics lessons began in the third year and "Dafty" astonished everyone by mastering geometry with no apparent effort. Promoted to a desk in the top group, he found himself in more sympathetic company and began to make friends. One of them was the star of the class, Lewis Campbell, and, by a stroke of luck, his family moved next door to Aunt Isabella's. On the walk home, the two boys began to share their thoughts on life, and James's world opened up. Now he had someone of his own age who would listen to his teeming ideas and counter with his own. They remained friends for life, and when Maxwell died, Campbell wrote a moving biography. At school, this friendship led to others, including one with Peter Guthrie Tait, who went on to become one of Scotland's great scientists.

At the age of fourteen, Maxwell published his first paper. It was on the kinds of curves that can be drawn with a pencil, a few pins and a piece of string. Most people know how to draw an ellipse using two pins and a simple loop of string, but by more complex looping, Maxwell produced whole families of curves. It wasn't unusual for him to produce geometrical propositions—he was doing it all the time—but his well-connected father decided to show this one to his friend Professor James Forbes at Edinburgh University, to see if anything like it had been done before. It turned out that the great French mathematician and philosopher René Descartes had investigated similar curves, but that Maxwell's construction was both simpler and more general. His paper was read out for him to the Royal Society of Edinburgh because he was deemed too young to do it himself. He had entered the scientific world of Edinburgh and met James Forbes, who was to play a big part in guiding his career.

All facets of young Maxwell's life shone brightly. Never one to bear a grudge, he now enjoyed the lively companionship of his schoolfellows and began to excel in English, history, geography, and French, as well as mathematics. He seemed to remember everything he had read and showed an amazing ability to write verse on any topic in impeccable rhyme and meter. He was less fluent in formal conversation—words came in spates between long pauses, and he was shy with strangers. His replies to questions tended to be indirect and enigmatic, leaving the questioner no wiser. Yet when at ease with friends, he reveled in the banter and would entertain them with a flow of surprising observations and metaphors on whatever was the subject of the moment. In the holidays at Glenlair, he joined in the Happy Valley social life, rode, walked the hills, helped with the harvest, and, in the winter, skated and took a hand at curling.

During the school term, his father visited Edinburgh whenever he could, and the two would take walks on Arthur's Seat, the rocky hill that overlooks the city, or visit other local attractions. Every new experience fed the boy's probing and retentive mind, and one in particular was eventually to bring a result that changed the world. John took James to an exhibition of "electromagnetic machines." These were early days, and the machines were devices like the magnetic beam engine, built for demonstration rather than for work, but all were testament to the discoveries of the great Michael Faraday. Young Maxwell had been introduced to the wonder of forces acting in space.

John Clerk Maxwell's plan was for his son to become a lawyer—a more successful one than he himself had been. This seems odd, given his son's obvious gift for science and his own fascination for technology, but his judgment was not wholly at fault this time. Science, then called natural philosophy, was generally thought to be an excellent hobby for a gentleman but a poor career choice: It was poorly paid and opportunities were sparse because there were few professional posts and the post-holders tended to remain for life, as Faraday did at the Royal Institution. Strange as it seems to us, science was not even thought to be particularly useful, as most of the great advances in industry and transport had been introduced and developed not by natural philosophers but by practical men with little theoretical knowledge, like Abraham Darby, the inventor of coke smelting, and George Stephenson, known as "the Father of railways." James himself had given little thought to the matter. He was drawn to science, but it was far from his only interest—literature and philosophy were stimulating, and perhaps the law would be just as compelling when he got to know it. He had learned enough to know how much more there was to learn, and he was keen to spread his wings. Edinburgh University was a fine place to start, and at the age of sixteen he enrolled there to complete his general education before studying for the law. By way of celebration, he wrote an ironic tribute to his old school in the form of a song—one that we might place somewhere between Robert Burns and Tom Lehrer. The first and last verses ran:

If ony hear has got an ear
He'd better tak a haud o' me
Or I'll begin, wi' roarin' din,
To cheer our old Academy

> Let scholars all, both grit and small
> Of learning mourn the sad demise;
> That's as they think, but we will drink
> Good luck to Scots Academies.[4]

The three years Maxwell spent at Edinburgh University are sometimes described as a fallow period when not much happened. In fact, they did much to make him the kind of scientist he was.

Scottish universities, Edinburgh in particular, had been inspired by the Enlightenment of the 1700s and early 1800s. They provided a broad education and strove to produce confident, well-rounded young men who could hold their own in any company. Philosophy held pride of place among the faculties: Edinburgh, home of the great David Hume, had two professorial posts, both occupied by famous men. John Wilson, celebrated under his pen name Christopher North, was professor of moral philosophy, and Sir William Hamilton, not to be confused with the Irish mathematician of the same name, was professor of mental philosophy. There was also the chair of natural philosophy, which we would now call science. This was held by James Forbes, who had been instrumental in getting fourteen-year-old Maxwell's paper on oval curves published by the Royal Society of Edinburgh.

Maxwell chose to study moral philosophy under Wilson, logic and metaphysics under Hamilton, science under Forbes, mathematics under Philip Kelland, and chemistry under Professor Gregory. There were some disappointments, the greatest being Wilson's lectures on moral philosophy, which, to Maxwell, only served to demonstrate that woolly thinking leads to wrong conclusions. Kelland's course on mathematics was, at first, too elementary to be interesting, though things improved later, and Gregory's lectures on chemistry suffered by being purely theoretical—laboratory experiments were supervised separately by his assistant, known as "Kemp the practical," who used quite different methods from those described by Gregory in the classroom. But even negative experiences were put to use: this one helped to form Maxwell's conviction that practical work must be an integral part of any science course, not a tacked-on extra.

The disappointments were offset elsewhere: Hamilton and Forbes were inspirational. William Hamilton's style was to instill in his pupils a spirit of relentless questioning and criticism. He had been instrumental in introducing the work of Immanuel Kant into Britain, and he used to stress Kant's proposition that nothing can be known about any object except by its relation to other objects. David Hume's notion of skepticism also played a large part in Hamilton's teaching: nothing can be proved, except in mathematics, and much of what we take to be fact is merely conjecture. Deep waters, but to Maxwell they were new and exciting, especially when Hamilton responded to his awkward questions by posing still-deeper questions.

Maxwell's study of philosophy served him well. One of the exercises he wrote for Hamilton gives us a glimpse of his ability to explore regions of scientific thought beyond the range of his fellow scientists.

> Now the only thing which can be directly perceived by the senses is Force, to which may be reduced light, heat, electricity, sound, and all the other things that may be perceived by the senses.

He had seen the truth of Kant's argument that the way we detect a solid object is by the *force* that resists our attempts to move through it. Twenty years later, when checking a draft of William Thomson and Peter Guthrie Tait's *Treatise on Natural Philosophy*, he had to put them right on this very point. They had defined mass wrongly and had to be told that "matter is never perceived by the senses."

Another characteristic of Maxwell's work, no doubt strengthened by his study of philosophy, was the way he could give full rein to his imagination, using the most surprising analogies, yet at the same time apply strict skepticism to his own results, even when they were brilliantly successful. This way, he was often able to return to a subject after a long interval and take it to new heights using a completely different approach.

James Forbes inspired Maxwell in a different and still more profound way. He was a true mentor, and their relationship stands comparison with that between Michael Faraday and Humphry Davy. Forbes's special passion was for Earth science—he pioneered the study of glaciers—but he was a consummate all-round scientist whose fascination with the physical world bound Maxwell in its spell. The two developed a rare rapport, and Forbes would let his pupil stay long after hours in the laboratory, making whatever investigation took his fancy. He could also be a hard task-master when required. When Maxwell submitted a poorly drafted paper to the Royal Society of Edinburgh and a colleague was asked to referee it, Forbes chose to deliver the sharp reproof himself. This was an act of kindness, and Maxwell knew it. Maxwell went on to develop the distinctive and crystal-clear writing style so admired by scholars—the beautifully written papers express not only his scientific ideas but his love for the literary tradition of the English language.

In a book review written many years later for the journal *Nature*, Maxwell gave what is clearly a sketch of Forbes.

> If a child has any latent talent for the study of nature, a visit to a real man of science in his laboratory may be a turning point in his life. He may not understand a word of what the man of science says to explain his operations, but he sees the operations themselves, and the pains and patience which are bestowed on them; and when they fail he sees how the man of science, instead of getting angry, searches for the cause of failure among the condition of the operation.

When his mentor died in 1868, Maxwell told a friend "I loved James Forbes."

Inspiration from Hamilton and Forbes was one of three factors from Maxwell's Edinburgh years that helped to form his character as a scientist. The second was the enormous amount of reading he did on all manner of topics—far more than most people accomplish in a lifetime. And he didn't just read; he analyzed, appraised, and remembered. This meant that he always had a vast store of knowledge to draw on for comparisons and analogies. The third factor, and the most important, was the free-wheeling experimenting he carried out during the holidays in an improvised workshop-cum-laboratory that he set up in a little room over the washhouse at Glenlair. He described it in a letter to Lewis Campbell.

> I have an old door set on two barrels, and two chairs, one of which is safe, and a skylight above, which will slide up and down.
>
> On the door (or table) there is a lot of bowls, jugs, jam pigs [jars], etc., containing water, salt, soda, sulphuric acid, blue vitriol, plumbago ore; also broken glass, iron, and copper wire, copper and zinc plate, bees' wax, sealing wax, clay, rosin, charcoal, a lens, a Smee's Galvanic apparatus [an electrical kit], and a countless variety of little beetles, spiders and woodlice, which fall into the different liquids and poison themselves.

He tried all the chemical experiments possible within his resources and, when visited by the estate children, let them spit on a mixture of two white powders to turn it green. Among countless other experiments, he copper-plated jam jars and made simple electromagnetic devices, including a model telegraph. But the most important equipment in the laboratory turned out to be the pieces of broken glass.

He had heard that if you shone plane polarized light through glass that was under strain, you would see colored patterns, so he set out to investigate. He cut the bits of broken window glass into geometric shapes, heated them to red heat, and cooled them rapidly, so that the outer parts would cool faster than the inner, leaving internal strains "frozen" into the glass. To get polarized light from ordinary sunlight, he made a polarizing apparatus from a large matchbox, a small sheet of mica, and two pieces of glass, cut to shape and stuck in with sealing wax at the correct angles. Results exceeded expectations. Each of the geometrically shaped specimens revealed its own beautiful pattern of colored lines. They mapped out perfectly the state of strain in the glass because each of the colored lines was a "contour," connecting points of equal strain.

To record his findings, he improvised a camera lucida, which made a virtual image of the patterns appear on a piece of paper so that he could copy them in watercolor. He sent the results to William Nicol, the famous Edinburgh optician. Nicol was so impressed that he sent Maxwell a pair of his prized Iceland spar polarizing prisms, the very kind that Faraday had used when detecting the effect of magnetism on light. Now that he had a readymade source of polarized light, Maxwell extended his investigation by passing the light through various jellies made using gelatin from the kitchen. He twisted the jellies to put them under torsional stress, observed the patterns of strain revealed by the colored lines, and copied them in watercolor and crayon. This was an early demonstration of the technique of photoelasticity that became tremendously useful to structural engineers. When they made a scale model of, say, their bridge, in a transparent material, put it under various loads, and shone polarized light through it, the patterns of strain would show up as colored lines, indicating any weak points where the structure might need strengthening.

All these DIY adventures not only honed Maxwell's experimental skill but also gave him a commanding insight into nature's processes and helped develop his uncanny intuition—Maxwell's guesses always seemed to be right. And along with the experimental work went his mathematical propositions, or "props" as he called them. Two were published by the Royal Society of Edinburgh. One was geometrical, on the paths traced out by a point on one kind of curve when rolled on another. The other was a remarkable achievement for a nineteen-year-old working almost entirely on his own. Called "On the Equilibrium of Elastic Solids," it was a theoretical accompaniment to his experimental work with polarized light and gave the full mathematical theory of photoelasticity based on strain functions. Both papers were read out for him as he was still too young to be allowed to do it himself.

Life was full, but Maxwell was missing the stimulating company of his close friends who had by now left Edinburgh University for Oxford or Cambridge. He filled long letters to Lewis Campbell with lighthearted accounts of his thoughts and researches at Glenlair, but it was no substitute for conversation and he implored his friend to visit. Campbell was at Oxford, but P. G. Tait and another friend, Allan Stewart, had gone to Cambridge. He began to worry that a matchless opportunity was slipping by. Cambridge was the place for aspiring scientists, and Forbes tried to persuade John Clerk Maxwell that James should go there too. John was reluctant—he would see less of his son as the university terms in England were longer, and there was the danger that James might fall under the influence of undesirable rich English students with their dissipated way of life. The decision was repeatedly put off and James was bracing himself to prepare for life at the Scottish Bar when events took a happy turn. He wrote to Campbell:

> I have notions of reading the whole of *Corpus Juris* and Pandects in no time at all; but these are getting dim as the Cambridge scheme has been howked up from its repose in the region of abortions, and is as far forward as an inspection of the Cambridge *Calendar* and a communication with Cantabs.

His father had at last agreed that he should go to Cambridge. Forbes was delighted and took the unusual step of writing to William Whewell, master of Trinity College, to tell him what to expect. (This was the same man whom Faraday had consulted on terminology.) Forbes wrote of Maxwell,

> He is not a little uncouth in manners, but withal one of the most original young men I have met. . . . He is a singular lad, and shy [but] very clever and persevering. . . . I am aware of his exceeding uncouthness, as well mathematical as in other respects. . . . I thought the Society and Drill of Cambridge the only chance of taming him and much advised his going . . . I should think he might be a discoverer.

Uncouth? Forbes was probably indulging in some donnish argot with Whewell, whom he knew well. Certainly Maxwell lacked sophistication; his manners did not match those of the polished products of Eton and Harrow that made up much of the Cambridge student population, and his accent was strange to the English ear. Even within the family he was censured for his want of social poise—when, at dinner, his attention was drawn to an interesting reflection in a glass or a swaying of a candle flame, Aunt Jane would recall him to the company with a sharp "Jamesie, you're in a prop." But he was strikingly good-looking and always neatly, though not fashionably, dressed—he hated starched collars and wore no gloves, even in the winter. He cared nothing for any kind of luxury, even traveling third class on the railway, saying that he preferred a hard seat. Lewis Campbell's mother summed him up very well:

> His manners are very peculiar; but having good sense, sterling worth, and good humour, the intercourse with a college will rub off his oddities. I doubt not of his being a distinguished man.

At nineteen he was already an experienced scientific experimenter who had accumulated a vast amount of knowledge and had published three mathematical papers. Yet he had never worked under the slightest pressure; there was tremendous power in reserve. In the autumn of 1850, he packed his Nicol prisms and as much as he could manage of his experimental paraphernalia into a trunk. The society and drill of Cambridge were about to do their work.

CHAPTER NINE

SOCIETY AND DRILL

1850–1854

After reporting to his tutor in St. Peter's College, known as Peterhouse, Maxwell was delighted to be given rooms with good light, just what he needed for his optical experiments. At peace with the world, he invited his old school friend Tait in for tea and they caught up with a long chat. The next day, there was a tour of the colleges, including homage at the tombs of Isaac Newton and Francis Bacon in Trinity College Chapel. The aura of scholarship seemed to be everywhere. But the lectures were a great disappointment at first—he found himself "spelling out Euclid" and "monotonously parsing a Greek play."* And his fellow students at Peterhouse seemed to be a snobby crowd, unreceptive to his attempts at genial discourse. His elation began to fade, and beneath it was an undercurrent of unease and restlessness. But all turned out for the best, thanks in large part to quite another factor.

His father had chosen Peterhouse, a small and elite college, but was now having second thoughts. An assiduous networker, he had discovered that one of James's fellow students, E. J. Routh, was a formidable mathematician who was likely to take one of the rare Peterhouse fellowships. Other colleges were better endowed and had more fellowships. The outcome was that after a term Maxwell moved to Trinity, the large and sociable college that Forbes had recommended all along.

Life at Trinity was a joy. Under William Whewell's aegis, the college had become a fertile ground for ideas and debate on just about any topic. Maxwell was in his element, and in a short time had a troop of friends. He joined in any discussion that was going on and relished the camaraderie and badinage. On some evenings he walked about looking for someone to bandy ideas with, and he met others doing the same. He acted as a kind of general factotum to the other students, nursing those who were sick and bucking up those who were depressed. When a friend had eye trouble and couldn't read, Maxwell spent an hour each evening reading out the man's bookwork for the next day. Meanwhile, his private reading went on apace and, when time allowed, he tinkered with his odd miscellany of scientific apparatus, getting ideas for proper investigation later. He even managed to write two scientific articles, one of which reported his remarkable proposal for a flat lens with a variable refractive index that would give perfect imaging. It became known as the fish-eye lens, and was reputedly inspired by close examination of a breakfast kipper.

It wasn't easy to fit all this in along with lectures and the work set by the lecturers and tutors, so he tried unusual daily routines—even jogging in the middle of the night to make sure he got enough exercise. A fellow student reports:

> From 2 to 2.30 am he took exercise by running along the upper corridor, down the stairs, along the lower corridor, then up the stairs, and so on until the inhabitants of the rooms along his track got up and lay *perdus* behind their sporting doors to have shots at him with boots, hair-brushes, etc., as he passed.[1]

Not all of Maxwell's experiments worked!

He gave full vent to his poetic muse, writing everything from translations of classic Greek odes to irreverent trifles dashed off to amuse his friends. In one of these he takes the part of the hypothetical "rigid body," beloved of lecturers in applied mathematics, in a parody of Burns's *Comin' through the Rye*.

> Gin a body meet a body
>   Flyin' through the air
> Gin a body hit a body,
>   Will it fly? And Where?
> Ilka impact has its measure

Ne'er a ane hae I,
But all the lads they measure me,
Or, at least, they try.

Gin a body meet a body
Altogether free,
How they travel afterwards
We do not always see,
Ilka problem has its method
By analytics high;
For me, I ken na ane o' them,
But what the waur am I?

He was invited to join an elite discussion group called the Apostles. Initially founded by twelve students as a secret society in 1820, the group had perpetuated itself by electing a new member to replace each man who left. At first the society was a forum for discussing progressive ideas on religion, politics, and education that had been resolutely ignored by the university. By Maxwell's time, thanks in part to the Apostles themselves, the university had become much freer, but the society was still a breeding ground for ideas that transcended conventional ways of thinking. Over the years, it has included among its ranks many who went on to make outstanding contributions to human thought and its expression—for example: Alfred Lord Tennyson, Rupert Brooke, Bertrand Russell, Ludwig Wittgenstein, Lytton Strachey, the theologian and social reformer F. D. Maurice, the mathematician G. H. Hardy, E. M. Forster, and John Maynard Keynes.

Meetings were traditionally held on Saturday evenings, when one member would read an essay on a preannounced topic which he would then throw open for discussion. Maxwell's contributions show that his thoughts ranged far beyond mathematics—he was, in the language of Plato, "taking a survey of the universe of things." It was a wonderful opportunity to present his brimming ideas to people with creative minds who would respond with ideas of their own, and he took it to the full. Among his essays were "Is Autobiography Possible?" "Has Everything Beautiful in Art Its Original in Nature?"

"Morality; Language and Speculation," and "Are There Any Real Analogies in Nature?"

Maxwell's studies of philosophy came to the fore in his "Analogies" essay, which, some say, holds the key to his seminal thinking in theoretical physics. Certainly, if one wants insight into how Maxwell was able to make progress where others were becalmed, this is a good place to look. At the heart of the essay was the Kantian philosophy that all human knowledge is of relations rather than things. As he put it:

> Whenever [men] . . . see a relation between two things they know well and think there must be a similar relation between things less known, they reason from one to the other. This supposes that although pairs of things may differ widely from each other, the relation in the one pair may be the same as that in the other. Now, as in a scientific point of view, the relation is the most important thing to know, a knowledge of one thing leads us a long way towards a knowledge of the other.

But not, of course, *all* the way—analogy was simply an aid to understanding, and he warned of the dangers of confusing it with identity. One needs to explore issues from all sides, as he made clear in a passage that provides us with a fine forty-nine-word summary of his scientific philosophy:

> The dimmed outlines of phenomenal things all merge into one another unless we put on the focusing glass of theory, and screw it up sometimes to one pitch of definition and sometimes to another, so as to see down into different depths through the great millstone of the world.[2]

A fellow member of the Apostles remembered that Maxwell always took part animatedly in the conversations, but that his tendency to speak in parables and his strange accent made it hard sometimes to get his meaning. It wasn't easy for newcomers to tune in to Maxwell's wavelength—to use an appropriate metaphor—but the effort was worth it. One of his fellow students remarked:

> Maxwell as usual is showing himself acquainted with every subject upon which the conversation turned. I never met a man like him. I do believe there is not a single subject on which he cannot talk, and talk well too, displaying the most out of the way information.[3]

At Cambridge, the bachelor's degree course took four years and was known as the Tripos, reputedly after the type of three-legged stool that candidates used to sit on when taking oral exams. By Maxwell's time, oral exams had given way to written ones and there were several of these at stages during the course, culminating in the fearsome Mathematical Tripos exam in the final year, which everybody had to pass to get a degree of any kind. The Mathematical Tripos had become an important institution in nineteenth-century Britain and served as a model for the introduction of competitive examinations throughout the country.

The exam took place in chilly January, and its setting was the Senate House, a building in the style of a Greek temple that had no fires or stoves. Students arrived in greatcoats and mufflers, sometimes to find their ink frozen in the inkwells. Every man had to take the first three days of exams, doing his best to solve problems against the clock for five and a half hours each day. Those who wanted an honors degree, even classics students, then came back for another four days of exams that were even more difficult. Those who gained first-class honors were awarded the title of wrangler, which brought lifelong recognition and gave a substantial boost to a career in any field. Wranglers were ranked in order, and to become senior wrangler was like winning an Olympic gold medal. The Mathematical Tripos was, indeed, treated by the press like a national sporting event, and large sums were bet on the outcome. After it, the best students sat an even harder set of papers for the Smith's Prize.

Mathematics had a long and illustrious history at Cambridge, starting from when Isaac Newton held the Lucasian Chair of Mathematics. Cambridge was the place where one studied mathematics, while Oxford's genius lay more in the humanities. From the mid-1700s onward, Cambridge had developed its highly competitive examination culture, in which highly coached students strove to be ranked as high as possible in the Tripos listing. The system produced not only some of the best British mathematicians and scientists of the nineteenth century but also eminent churchmen, doctors, lawyers, civil servants, and economists. Success in the high-pressure Tripos was held to be a sign of a first-rate mind, able to work under trying conditions and solve problems in any field. Oxford, by contrast, largely eschewed competition, perhaps thinking it an unseemly activity for gentlemen; honors classifications there were never individually ranked—nor were all students expected to study mathematics.

The Tripos required virtuosity in solving the set problems quickly and accurately. These problems usually had little connection with reality—they were contrived puzzles that demanded mastery of a wide repertoire of tricks and shortcuts. This was not Maxwell's forte, so he set out to master these skills. It was routine for all students aiming at high honors to be coached by one of the freelance private tutors whose income depended on results, and Maxwell joined the class of William Hopkins, the renowned "wrangler maker." Hopkins was a driver but not a dull one, as one of his old students, the anthropologist Francis Galton, recalled:

> Hopkins to use a Cantab expression is a regular brick, tells funny stories connected with different problems, and is in no way Donnish, he rattles at a splendid pace and makes mathematics anything but a dry subject. I never enjoyed anything so much before.

Hopkins had coached more than two hundred wranglers[4] but had never met anyone like Maxwell, whom he described as "unquestionably the most extraordinary man I have ever met with in the whole of my experience." "It is not possible," he said, "for Maxwell to think incorrectly on physical subjects; in his analysis, however he is far more deficient." It was Hopkins's job to subject Maxwell's freewheeling mind to what Forbes had called the "drill" of Cambridge. P. G. Tait, who had also studied under Hopkins, described the task that faced the tutor, and the manner in which his friend responded to drilling.

> [Maxwell] brought to Cambridge in the autumn of 1850 a mass of knowledge which was really immense for so young a man but in a state of order appalling to his methodical private tutor. Though the tutor was William Hopkins, the pupil to a great extent took his own way, and it may safely be said that no high wrangler of recent years ever entered the Senate House more imperfectly trained to produce "paying" work than did Clerk Maxwell.

Maxwell did, however, learn much from Hopkins—he came to see the advantages of attacking problems systematically and the usefulness of the standard algebraic procedures. By applying routine drills and checks he was able to reduce his tendency to make algebraic mistakes, though this remained a weakness—"I am quite capable of writing a fancy formula," he used to say, meaning a wrong one. Like Faraday, he liked to picture a problem, and on at least one occasion when Hopkins had filled the blackboard with equations, he solved the problem in a few lines with a diagram. Though he never "crammed" in the way that was thought necessary for high honors in the Tripos, he dutifully carried out all the set work. A fellow student, W. N. Lawson, reports:

> Maxwell was, I daresay you remember, very fond of a talk upon almost anything. He and I were pupils (at an enormous distance apart) of Hopkins, and I well recollect how, when I had been working all the night before and all the morning at Hopkins' problems with little or no result, Maxwell would come in for a gossip and talk on and on while I was wishing him far away, till at last about half an hour before our meeting at Hopkins', he would say—"Well, I must go to old Hop's problems"; and by the time we met they were all done.

A few months before the Tripos, Maxwell became an observer in a widely debated university controversy. F. D. Maurice, a former student at Trinity and member of the Apostles, had gone on after leaving Cambridge to found the Christian Socialist movement, which had attracted a following among Maxwell's fellow students. The movement aimed to counteract the dehumanizing effects of industrial work by setting up cooperatives and working-men's colleges. Maurice's *Theological Essays* had caused a stir in 1853 by seeming to question some of the Articles of the Church of England, and as a result he was summarily fired from his professorship at King's College, London. Maxwell, along with some of his friends, was appalled at Maurice's treatment, especially since he firmly supported the idea of education for the working man. Maxwell had already helped young farmworkers at Glenlair by lending them books from the family library, and later, as a fellow at Cambridge; and as a professor at Aberdeen and at King's College, London, he gave up an evening a week to teach at working-men's colleges. He clearly felt, along with Faraday, that the common man was capable of appreciating science and had a right to know about it.

He didn't let up in his non-Tripos activities and, perhaps, overdid things. On a vacation with a friend's family in Suffolk, he developed a fever and became delirious. The family nursed him for two weeks and sent daily reports to his father. Much moved, and profoundly grateful for their kindness, Maxwell nevertheless couldn't help making a critical observation on their way of life. Everyone was so anxious to know and to heed everyone else's wishes that no one had a life of his own. How much better it would be, Maxwell thought, for everyone to be allowed some interests that were not to be too much encouraged or inquired into by the others—in his view, this would greatly increase the resources of the family as a whole.

Exam time came, and he sat with the others day after day in the Senate House, at his father's suggestion wrapping his feet in a bit of blanket for warmth. In the evenings they all needed to relax, and a merry crowd gathered in Maxwell's rooms to dabble in experiments with magnets under his guidance. The results soon followed. E. J. Routh was announced as senior wrangler and J. Clerk Maxwell was placed second. The best students then competed for the Smith's Prize, and here Routh and Maxwell were declared joint winners. Routh was an exceptional mathematician who went on to do outstanding research. He did much to systematize the mathematical theory of mechanics; he created several ideas that found application in modern control theory; and, like the giants Laplace, Lagrange, and Hamilton, he gained the rare distinction of having a function named after him, the Routhian.[5] He also took up coaching others for the Tripos and became the supreme "wrangler maker," surpassing even Hopkins. Maxwell had done well, not quite as well as his friend Tait, who had been senior wrangler two years earlier, but Tait had not been up against Routh. He had established his mathematical credentials and put himself in a good position to gain a fellowship at Trinity. His father was delighted at the results, and congratulations tumbled in from uncles, aunts, and cousins.

Maxwell's undergraduate years were joyful ones. And productive—he had completed the Tripos grind with honor and, thanks to Hopkins's drilling, his mathematics now had the discipline and poise that Forbes had earlier brought to Maxwell's experimental work. Society had played its part, too, by softening his eccentricities: he now impressed strangers as an interesting young man rather than merely an odd one. Best of all, he had made a set of lifelong friends. They included H. M. Butler, who went on to be headmaster of Harrow School and then master of Trinity College; F. W. Farrar, who became dean of Canterbury but is best remembered as the author of the popular moral tale *Eric, or Little by Little*; and R. B. Litchfield, who founded the London Working Men's College. The value that Maxwell put on friendship is evident from a letter he wrote later to Litchfield when another friend, Robert Henry Pomeroy, died in the Indian Rebellion of 1857.

> It is in personal union with my friends that I hope to escape the despair which belongs to the contemplation of the outward aspect of things with human eyes. Either be a machine and see nothing but "phenomena" or else try to be a man, feeling yourself interwoven, as it is, with many others and strengthened by them whether in life or death.

Lewis Campbell gives us a picture of Maxwell as his friends saw him:

> His presence had by this time fully acquired the unspeakable charm for all who knew him which made him insensibly become the center of any circle, large or small, consisting of his friends or kindred.

The society of Cambridge had done him good, as Forbes had predicted, but Maxwell had more than repaid the debt. A fellow student told Campbell:

> Of Maxwell's geniality and kindness of heart you will have had many instances. Everyone who knew him at Trinity can recall some kindness or some act of his which left an ineffaceable impression on the memory—for "good" Maxwell was in the best sense of the word.[6]

After consulting his father, Maxwell decided to stay at Trinity as a bachelor scholar and apply for a fellowship. This wasn't a long-term plan—in those days, fellows of Trinity were required to be ordained into the Church of England within seven years and to remain unmarried, and he had no intention of making either commitment—but it did mean that he could now take up the ideas for scientific investigations that had been brewing in the back of his mind. He wanted to come to grips with electricity, but at this time the most pressing question on his mind was, how do we see colors? The way he answered it gives us an insight into his boldness, ingenuity, and resolution.

CHAPTER TEN

AN IMAGINARY FLUID

1854–1856

When Maxwell was three, someone had said, "look at that lovely blue stone," and he had responded, "but how d'ye *know* it's blue?"[1] The question had still not been answered. In the early 1800s, Thomas Young had put forward the interesting idea that the human eye has three types of receptors, each sensitive to a particular color, and that the brain combines the signals to form a single *perceived* color. But Young couldn't supply any supporting evidence and his theory had been largely neglected for the best part of half a century when James Forbes thought of taking a disc with differently colored sectors, like a pie chart, and spinning it fast so that one sees not the individual colors but a blurred-out mix. His idea was that each of the eye's three types of receptors might respond to one of the primary colors used by artists—red, yellow, and blue—so he tried various mixtures of these colors on the spinning disc to see what combined color would appear. The results were puzzling. When, for example, he mixed just yellow and blue, he didn't get green, as painters did, but a dull sort of pink. And he couldn't get white, no matter how he mixed the colors.

This was as far as Forbes got, but Maxwell took up the idea and soon discovered the source of his mentor's confusion. Forbes had failed to distinguish between mixing colors in the light that reaches the eye, as when spinning a disc, and mixing pigments, as a painter does. Pigments *extract* color from light—what you see is whatever light is left over after the pigments have done their extraction. So perhaps Forbes's choice of the painter's primary colors, red, yellow, and blue, was wrong. Maxwell tried, instead, mixing red, *green*, and blue, and the outcome was spectacular. Not only did he get white by using equal proportions of the three colors, but he found that he could produce a great variety of colors, also, simply by varying the proportions of red, green, and blue.

To put things on a proper footing, he ordered sheets of colored paper in many colors from an Edinburgh printer and had a special disc made—he called it his "color top." About six inches in diameter, it had percentage markings around the rim, a handle, and a shank for winding a pull-string. He cut out paper discs of red, green, and blue and slit them so that they could be overlapped on the color top with any desired amount of each color showing. This way, he was able to measure what percentages of red, green, and blue on the spinning disc matched the color of whatever paper was held alongside the spinning color top for comparison. But then he thought of a better arrangement. Instead of holding a separate piece of paper alongside, he put a smaller paper disc of the color he wanted to match on top of the red, green, and blue ones, so that it occupied the inner part of the top's disc. With this arrangement, he could add a sector of black, if needed, to be able to match brightness as well as hue.

Using this homely device, he showed that you really can get any color you want by mixing red, green, and blue in the right proportions—exactly the principle used in our television sets today. Maxwell got all of his friends and colleagues to have a go at mixing colors, and he found remarkably little variation in color perception among people with normal vision. He particularly sought out color-blind people and found that most of them lacked fully functioning red-sensitive receptors, which explained their difficulty in telling red from green.

All of this was groundbreaking work, but there was no fanfare: Maxwell simply sent off a paper to the Royal Society of Edinburgh and demonstrated some of the results to the Cambridge Philosophical Society using his color top. In his own view, the work done so far was no more than a preliminary sketch because the colors of the printer's paper were arbitrary—simply "specimens of different kinds of paint."[2] To get precise and replicable results, he needed to use pure spectral colors extracted from sunlight, so he devised a "color box" to do this, using a prism to spread the sunlight into a spectrum, adjustable slits to select particular colors, and further optical arrangements to combine the selected colors. Over the years, he made several versions of the color box, improving it each time, and the work became a lifelong project. Had he done nothing else, he would now be known as Maxwell, one of the great founders of the science of color vision.[3]

While spinning the top, he had been fulfilling his duties as a bachelor scholar, giving classes and supervising examinations. The role wasn't onerous, but he took it seriously and volunteered to take extra classes, which was good practice for a professorship later on. He strongly supported the movement led by F. D. Maurice for workingmen's colleges and began the practice that he continued in Aberdeen and London of giving up an evening a week to talk at the local college. Friendships multiplied, and his social life was full. He continued his association with the Apostles and was elected to the exclusive Ray Club, a forum for discussing and promoting the natural sciences. As for exercise, there was walking, rowing on the Cam, jumping and vaulting in the gymnasium, and swimming in the new pool, where he helped to organize group sessions to make things more sociable. As if this were not enough, he kept up his formidable regime of general reading, taking in, among others, Carlyle, Chaucer, Francis Bacon, Pope, Goldsmith, Berkeley, and Cowper.

On a parallel track, his thoughts turned more and more to electricity and magnetism. Years of lighthearted experimenting—copper-plating jam jars, playing with magnets, building model telegraphs—had given him a fascination for the subject, and the time had come to begin a serious study. However, it wasn't at all clear where to start; he needed advice, and, through a family connection, he knew exactly where to find it.

His cousin Jemima had married Hugh Blackburn, a professor of mathematics at Glasgow University. The professor of natural philosophy there, and Blackburn's closest friend, was none other than William Thomson, who, as we have seen, was one of the few scientists to have taken Faraday's idea of lines of force seriously. Maxwell had met Thomson several years earlier when visiting Jemima with his father. Like everyone, he was impressed by the dashing young man of science, and the feeling was mutual. Maxwell wrote to Thomson from Cambridge, breezily announcing his intention "to poach on your electrical preserves" and asking for a reading list. Thomson was delighted to take on the role of mentor—he had many other interests and was now becoming involved in the great Atlantic telegraph-cable project.[4] His reply to Maxwell hasn't survived, but we can be sure that Faraday's *Experimental Researches in Electricity* had a prominent place on the reading list.

Scanning the books and papers that Thomson had recommended, Maxwell soon saw that the state of knowledge about electricity and magnetism was unsatisfactory. Much had been written, but each leading author had his own methods, terminology, and point of view. All the theories except Faraday's were mathematical and based on the idea of action at a distance. Their authors had largely spurned Faraday's notion of lines of force because it couldn't be expressed in mathematical terms, except, in a limited way, through an analogy Thomson had made between electric lines of force and the steady flow of heat through a metal bar.

Nevertheless, Maxwell was drawn to Faraday, both by Thomson's encouragement and by his own intuition—truth lay in observed results, and anyone aspiring to solve the remaining mysteries of electricity and magnetism should first study what had been found by experiment. He resolved to read all of Faraday's *Researches* before tackling the mathematical treatments. He was struck at once by Faraday's openness and integrity, and, as he read more, he came to see the intellectual strength of the work. Maxwell's experience from all the hours spent working in his improvised laboratory at Glenlair made him marvel not only at the precision of the great man's experiments but also at the power and subtlety of the reasoning that followed. To Maxwell, Faraday's ideas rang true: he had found a kindred spirit and a new source of inspiration. The warmth he already felt for Faraday is evident from a comment he later included in his *Treatise on Electricity and Magnetism*:

> The method which Faraday employed in his researches consisted in a constant appeal to experiment as a means of testing the truth of his ideas, and a constant cultivation of ideas under the direct

> influence of experiment. . . . Faraday . . . shows us his unsuccessful as well as his successful experiments and his crude ideas as well as his developed ones, and the reader, however inferior to him in inductive power, feels sympathy even more than admiration, and is tempted to believe that, if he had the opportunity, he too would be a discoverer.[5]

Having read Faraday, then Ampère, he turned indefatigably to the other authors and sent a progress report to Lewis Campbell:

> I am working away at Electricity again and have been working my way into the views of the heavy German writers. It takes a long time to reduce to order all the notions one gets from those men but I hope to see my way through the subject and arrive at something intelligible in the way of a theory.

He did arrive at a theory—though the task took three stages spread over nine years—and it was based on Faraday's concept of lines of force in space. He had already seen that mathematical writers were quite wrong to dismiss lines of force as an idle speculation or flight of fancy—the concept had evolved over years of solid experimentation and painstaking thought. The immediate task became clear—to find a way of expressing Faraday's ideas in mathematical language. By doing so, he hoped to demonstrate their equivalence to other theories, to confound Faraday's critics, and, with luck, to establish a base on which to build a fuller theory.

Thomson had opened the door a crack with his analogy between electric lines of force and heat flow. Maxwell now looked for a more general analogy, and he found one: the steady flow of an imaginary fluid through a porous medium. Amazingly, by this simple means, he was able to model all the known properties of static electric and magnetic fields, and to show that the relevant formulas could be derived equally well from assumption of action at a distance or from Faraday's lines of force.

Maxwell's fondness and flair for analogies were evident from the essay "Are There Any Real Analogies in Nature?" which he had written a few years earlier for the Apostles, and stemmed from his study of philosophy at Edinburgh. He remembered especially how Hamilton had stressed Immanuel Kant's proposition that all human knowledge is of relations rather than of things. Other physicists, less versed in philosophy, thought his use of analogies idiosyncratic, and few really understood what he was getting at. Some took him literally and couldn't see what the mechanics of fluid flow had to do with electricity and magnetism. But they should have heeded his clear warning to readers not to take his model to represent any kind of physical reality. It offered, he was at pains to stress, no physical theory of electricity or magnetism. He was simply attempting to "shew how, by a strict application of the ideas and methods of Faraday, the connection of the very different order of phenomena which he has discovered may be clearly placed before the mathematical mind."[6] The fluid itself was merely an aid to thought—"a collection of imaginary properties"—its purpose was to enable one to find the appropriate mathematical relationships without being committed to any particular physical theory.[7]

Maxwell's imaginary fluid was weightless, friction-free, and incompressible. This last property was the key to the analogy. It meant that the fluid had its own built-in inverse-square law: the speed of a particle of fluid flowing directly outward from a point source was inversely proportional to the square of its distance from the source. This, as he explained, was a just a matter of geometry. The amount of fluid that emerges per second from any sphere centered on the point source is the same, whatever the size of the sphere. So, as the sphere's surface area ($4\pi r^2$) is proportional to the square of its radius, the fluid must move outward at a speed inversely proportional to the square of its distance from the source. If the source is replaced by a sink, the same applies in reverse—this time the velocity is inward. And Maxwell showed that if there were any number of sources and sinks of any shape in any configuration, the speed and direction of fluid flow at any point could, in principle, be calculated by summing mathematically all the flows from each point on each source and to each point on each sink.

Electric and magnetic forces were known to follow a similar

law—the force between two electric charges or two magnetic poles was inversely proportional to the square of their distance apart—so the basis of the analogy was set. The direction and *speed of flow* of fluid at any point represented the direction and *strength* of either the electric force or the magnetic force; the faster the flow, the stronger the force. It was a strange analogy—moving fluid representing static force—but it served Maxwell's purpose. And the beauty of it all was that the streamlines of fluid flow represented Faraday's electric or magnetic lines of force.

Faraday had thought of the lines as discrete—he always talked of the *number* of lines—but Maxwell merged them into a continuous entity called flux. Electric or magnetic flux was the total amount of force that acted through any given cross section—one might think of it as akin, for example, to the amount of sunlight striking a particular patch of ground. In any small region in space, the flux had both a direction and a concentration, or density. A high density of flux corresponded to a high concentration of Faraday's discrete lines—the higher the flux density in that part of space, the stronger was the electric or magnetic force there. In Maxwell's analogy, the direction of his fluid flow in any part of space corresponded to direction of the electric or magnetic flux there, and the *speed of flow* corresponded to the flux density. To track the fluid's motion, Maxwell constructed imaginary tubes for it to flow along. The tubes behaved as though they had real walls because the lines of flow never crossed one another, and the whole system of tubes fitted together, leaving no gaps. The fluid traveled fast where the tubes were narrow and slower where they broadened out. Electric and magnetic flux were similarly contained in tubes; and, by analogy, forces were strong where the tubes were narrow and the flux dense, and weaker where the tubes were wide and the flux sparse.

The amount of fluid that flowed per second past any cross section of a tube was the same wherever the cross section was positioned. This rate of fluid flow corresponded to the amount of flux that acted through any cross section of the tube, and this quantity was also the same no matter where the cross section was placed. Maxwell defined a unit tube of flow as one that passed a single unit of volume of fluid per second, and, by analogy, a unit tube of flux as one that held a single unit of flux. A unit tube of flow was one that passed one milliliter of fluid per second, and the corresponding unit tube of flux held one unit of flux across any cross section throughout its length. Any quantity of flux could now be described as the relevant number of unit tubes—the calibration could be as fine as need be by making the unit of flux suitably small. Mathematical physicists could now interpret Faraday's "vague and varying" lines of force (to use Sir George Airy's description) as Maxwell's mathematically impeccable unit tubes of flux.[8] Perhaps to emphasize the point, Maxwell used the term "unit line of force" as an alternative term for a "unit tube of flux."

What made Maxwell's fluid move was pressure difference. Along each tube, fluid flowed from a relatively high-pressure source to a lower-pressure sink, the pressure falling as the fluid passed along the tube. In the electrical analogy, a source was a positively charged body at a relatively high electrical potential and a sink was a negatively charged body at a lower potential. Pressure difference in the fluid model represented the difference in electrical potential, which we now call voltage, and the rate of fluid flow along one of Maxwell's tubes represented electric flux. In any small region, the rate of fluid flow was proportional to the pressure gradient (the fall in pressure per unit distance) there, and, similarly, the concentration, or density, of electric flux[9] was proportional to the potential gradient (the fall in potential per unit of distance). Maxwell called this potential gradient the intensity, or simply the force, of the electric field.

In Maxwell's fluid model of the static electric field, substances like metals, in which electric currents could flow freely, took no part except that their surfaces could act as sources or sinks. Electric lines of force occurred in *insulators*—substances in which currents did not flow. As Faraday had found, these substances varied in their ability to conduct electric lines of force—each had its own specific inductive capacity. For example, glass conducted electric lines of force more readily than wood. In his model, Maxwell accommodated this property simply by endowing each substance with the appropriate amount of resistance to fluid flow—the lower the resistance, the smaller the pressure gradient necessary to produce a given speed of flow. By analogy, the greater the inductive capacity of the substance, the lower the potential gradient necessary to produce a given density of flux. A simple equation summed it up: The electric flux density at any point was equal to the electrical potential gradient there multiplied by the electrical inductive capacity of the substance. The analogy for static electric fields was complete.

Maxwell had done something remarkable. By using fluid pressure as an analogy for electrical potential, he had connected Faraday's concept of lines of electric force, regarded by most mathematical physicists as "vague and varying," with the abstract and precise concept of potential, which had come from mathematical astronomy. Pierre Simon Laplace had used the concept of *gravitational* potentials so successfully in his *Mécanique céleste* that it was natural for others to apply the same technique to electricity. Now Maxwell had given the mathematicians a direct route to Faraday's ideas, should they choose to take it.

The corresponding analogy for magnetic fields was a little more complicated. Maxwell began by considering a special case—the field surrounding a permanent iron magnet, such as the familiar bar magnet. This could be modeled in exactly the same way as a static electric field: a source and a sink became the north and south poles at the ends of the magnet; the pressure gradient at a point became the intensity, or force, of the magnetic field there; the resistance of the medium, or rather its reciprocal, became the magnetic inductive capacity; and the speed and direction of flow became the magnetic flux density. The fluid flow represented Faraday's magnetic lines of force, alias tubes of flux, and its pattern was exactly that revealed by iron filings sprinkled on a piece of paper held over a magnet. So far, so good, but this model fell short of what was needed—it did not explain why Oersted's compass needle had twitched when placed near an electric current.

The strange force that Oersted had discovered, acting at right angles to the current, was different from anything else encountered in nature. To tackle the problem, Maxwell drew inspiration from Faraday's longtime correspondent, Ampère. As we've seen, Ampère had found that a small loop of current acted like a magnet. Maxwell went further by showing that the magnetic effect of a large current loop, or circuit, was exactly the same as that of a magnetic shell. This shell was an imaginary surface bounded by the circuit, and the whole thing acted as a weird kind of magnet; the whole of one side of the surface was its north pole and the other side was its south pole, the strength of the poles being proportional to the current. As Maxwell explained, the shell worked because it could be thought of as many small loops of current, each acting as a magnet, put together in a mesh. In the mesh, all the internal currents canceled out because every part of every internal loop was shared with a neighboring loop and so carried equal and opposite currents that annulled one another. So the combined effect of all the small loops was exactly the same as that of the single large loop of current that ran around the edge of the mesh.

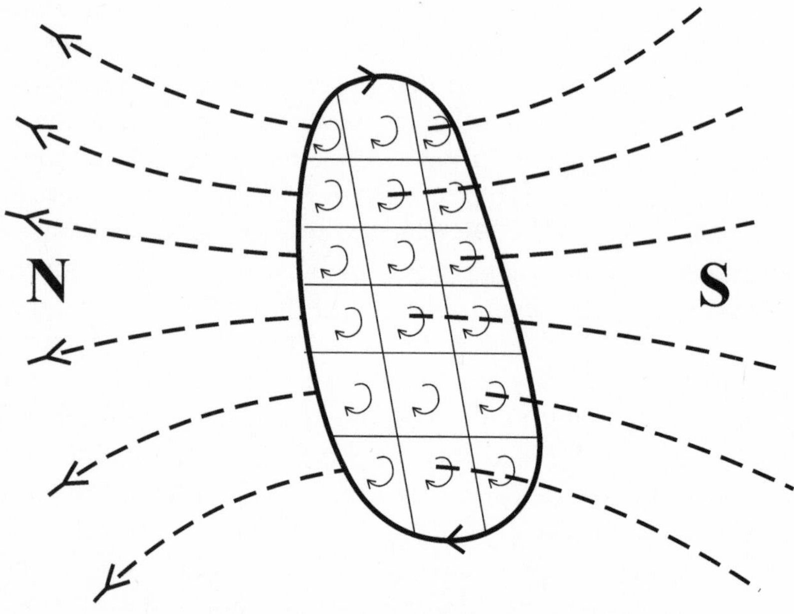

Fig. 10.1. Equivalence of a magnetic shell and a current loop. (Used with permission from Lee Bartrop.)

The magnetic effect of any current-carrying electric circuit could, therefore, be simulated by an imaginary magnetic shell, which would cause lines of force to circle from the north face of the shell around the circuit to the south face, forming the same pattern that iron filings take when sprinkled on paper over a magnet. They would "embrace" the circuit, to use Maxwell's word. Having completed a circuit, each line of force (or tube of flux) would join onto itself, forming a continuous loop.

In Maxwell's fluid analogy, the magnetic shell became a kind of pump that drove the fluid from its "north" face around the surrounding space to its "south" face, each streamline of flow joining onto itself. But the magnetic shell didn't have a fixed shape. In fact, it could take on any shape; the only constraint was that it was bounded by the current-carrying circuit. You could construct the shell in a way that would include any point on any of the streamlines of fluid flow. This meant that the pumping action took place all the way around every line of flow; it caused the fluid pressure to fall continuously all the way around, and this was what made the fluid move.

Here, the fluid analogy ran into a difficulty: you couldn't have continuously falling pressure all the way around a loop that joined onto itself. Maxwell didn't attempt any mechanical explanation, but he did explain what happened to the analogue of fluid pressure, the magnetic potential. It did, indeed, fall continuously as one went around the loop and was lower when one arrived back at the starting place than it had been before starting the journey. If one went round the loop the other way, against the direction of the analogous fluid flow, the potential was higher when one arrived back at the starting point than when one started, and if one repeated the process, the potential became higher and higher. So the potential at any point didn't have a single value; it depended on the number of times one had circled through the current loop. The difference in magnetic potential encountered in making one circuit around the current loop was the equal to the amount of mechanical work needed to move a unit magnetic pole around the loop. It represented the convertibility of electromagnetic energy into mechanical energy, and vice versa, that Faraday had discovered with his electric motor and generator.

Bypassing the limitation in his analogy with fluid pressure, Maxwell used the fluid model in the same way that he had done in the electrical case to produce a simple formula: the magnetic flux density at any point was equal to the magnetic force there multiplied by the medium's magnetic inductive capacity. None of his findings from the fluid model were really new—they could all be derived from the assumption of action at a distance. But he had presented them in a wholly new light and shown that the known formulas of static electricity and magnetism could be explained equally well by Faraday's lines of force as by action at a distance.

Maxwell extended the fluid analogy to model the flow of steady currents flowing through a resisting medium, but that was as far as it could take him. He had dealt with static electric and magnetic fields and steady currents, but the daunting task of working out the interactions of *changing* fields and currents lay ahead. It must have looked like a sheer cliff face, but there was more he could do even now. Faraday had conjectured that, even when things were static, a wire in a magnetic field was under a kind of strain that he called the electrotonic state. To Maxwell, this seemed a sound working hypothesis, and he signed off part 1 of his paper "On Faraday's Lines of Force" by signaling his intention to investigate the matter in part 2.

> I propose, in the following investigation to use symbols freely, and to take for granted the ordinary mathematical operations. By a careful study of the laws of elastic solids and of the motion of viscous fluids, I hope to discover a method of forming a mechanical conception of this electrotonic state adapted to general reasoning.

In part 1 he had stuck to his original plan to represent Faraday's ideas using only word descriptions and simple equations. Now he needed some power tools and looked into what had already been done in the mathematics of vectors, quantities that had both magnitude and direction. The principal authors here were the great German mathematician Carl Friedrich Gauss, the Englishman George Green, and the Scot William Thomson. Making good use of their work, Maxwell derived a set of equations connecting electric and magnetic fields. They represented everything that was known at the time, though, as Maxwell was to discover later, a vital part of the linkage between electricity and magnetism was still missing. He also found a mathematical expression for the electrotonic state but was not yet able to assign it a physical role. All this was expressed in very complicated-looking mathematics.

Though surpassed by his later writings, Maxwell's "On Faraday's Lines of Force"[10] is, surely, one of the finest examples of creative thought in the history of science. In his book *James Clerk Maxwell: Physicist and Natural Philosopher*, Francis Everitt shrewdly characterizes Faraday as a cumulative thinker, Thomson as an inspirational thinker, and Maxwell as an architectural thinker. Maxwell had not only found a way to express Faraday's ideas in mathematical language but also built a foundation for still-greater work yet to come.

For now, he had done all he could. Maxwell presented his paper to the Cambridge Philosophical Society in two parts spanning the 1855 Christmas holiday and consigned his thoughts on electricity and magnetism to what he called "the department of the mind conducted independently of consciousness."[11] He believed strongly in the power of subconscious thought to generate insights and, as he often did, expressed the idea in a poem.

> There are powers and thoughts within us, that we know not till
> they rise
> Through the stream of conscious action, from where self in
> secret lies.
> But when will and sense are silent, by the thoughts that come and
> go
> We may trace the rocks and eddies in the hidden depths
> below.[12]

Indeed, it would be six years before his next paper on electromagnetism appeared, and this set a pattern for the way he worked on his two great themes of electromagnetism and kinetic theory—he would write a paper during a few months of intense thought, then consign the topic back to the subconscious for several years (during which he wrote brilliant papers on other topics) before starting the next one. As we will see, much happened in the next six years, and what he produced then was a quite different set of analogies that represented changing fields as well as static ones and, in doing so, revealed what had been the missing part of the linkage between electricity and magnetism. They didn't know it, but his audience at the Cambridge Philosophical Society had witnessed the first of three stages in which Maxwell, inspired and guided by Faraday's ideas, created a theory of electromagnetism that would come to change our lives and to lay the foundations for twentieth-century developments in physics.

Life wasn't all work. One summer he enjoyed a memorable holiday in the Lake District, where he had been asked to join his uncle Robert's family. Robert Dundas Cay was his mother Frances's younger brother, and he had five children. Maxwell was very fond of all his Cay cousins, especially Lizzie, and this time they fell in love. She was only fourteen, but they planned to marry when she was sixteen; such marriages were not uncommon in those days. After the holiday, Maxwell walked the fifty miles home to Glenlair from the Carlisle railway station with a joyful heart, but the euphoria didn't last. Fearful of consanguinity, the family persuaded them to abandon any thought of getting married. It was a deeply wounding experience for them both, but they got on with their lives and eventually married other people.

In 1855, Maxwell was elected a fellow of Trinity College. He would now need to look for a professorship because the university required fellows to be ordained into the Church of England within seven years of election. There was no hurry, but a chance came sooner than expected. In January 1856, a letter came from James Forbes to tell him that the chair of natural philosophy at Marischal College, Aberdeen, was vacant, and to suggest that he apply. Maxwell discussed the matter at length with his father and mulled things over. It would be a jolt, after five happy years, to leave the center of the academic world for a distant northern outpost. On the other hand, he had observed the narrowing tendencies of college life at Cambridge, and it would be good to be out, feeling "the rubs of the world," as he put it to Campbell.[13] He would need to seek a job in the next few years anyway, and opportunities like this one didn't come up often. Besides, the shorter terms at a Scottish University would allow him to spend more time at Glenlair with his now-ailing father. He decided to apply. At first he had no idea how to go about it, but he soon learned that he needed to ask "swells" for testimonials. This presented no problem, but he was surprised to be asked to supply one himself for a less well-connected acquaintance who had applied for the same post.[14] He did. Another candidate was his old friend P. G. Tait, who was now professor of mathematics at Belfast University but wanted to return to Scotland.

John Clerk Maxwell was buoyed by the prospect of seeing more of his son. For a while, his health, which had been declining, seemed to improve, but it took a turn for the worse during the Easter vacation, and he died peacefully early one morning after Maxwell had nursed him through a troubled night. Maxwell had lost his closest companion—he and his father had been constantly in each other's thoughts and had written almost daily when apart—but sorrow was mixed with pride. He saw the love and respect that so many people had felt for his father, and he knew how fortunate he was to have had wise and loving parents. He wrote to relations and friends, organized the funeral, and assumed his father's mantle as laird, a role he took to heart.

To the casual observer, he would have appeared little changed after his father's death. He saw to estate business, kept up his correspondence, and carried on with his scientific pursuits. But inwardly he was devastated, and those close to him could see the pain beneath the surface. Lewis Campbell later recalled the great depth of feeling that underlay Maxwell's quiet demeanor at this time. As was his way at times of great joy or sadness, Maxwell expressed his emotions in a poem. Nothing could show more plainly the love of a son for his departed parents.

> I have leapt the bars of distance—left the life that late I led—
> I remember years and labors as a tale that I have read,
> Yet my heart is hot within me, for I feel the gentle power
> Of the spirits that still love me, waiting for this sacred hour

Yes—I know the forms that meet me, but are phantoms of the brain,
For they walk in mortal bodies, and they have not ceased from pain.
Oh! those signs of human weakness, left behind for ever now,
Dearer far to me than glories round a fancied seraph's brow.
Oh! the old familiar voices! Oh! the patient waiting eyes!
Let me live with them in dreamland, while the world in slumber lies.[15]

Back at Cambridge, he heard that his application had been successful. At the end of the summer term, he packed up his papers, color top, magnets, prisms, and the rest of his experimental paraphernalia, and, with fond memories, left Cambridge for Glenlair. It was the first time he had returned home without his father being there to welcome him. He spent much of the summer dealing with the estate, learning its detailed workings, and planning how he might fulfill his father's wish for more improvements when funds allowed. From time to time he entertained relations and Cambridge friends, but it was the loneliest episode in his life. In seven months, he left Glenlair only once; that was to make a short trip to Belfast to arrange for his cousin William Dyce Cay to study engineering under William Thomson's brother James. He managed to fit in a little scientific work and built a version of his color box rugged enough to withstand the journey to Aberdeen. The time for that journey approached, and part of his preparation was to draft what he called "a solemn manifesto to the Natural Philosophers of the North,"[16] his inaugural speech, something that was then expected of all new professors. In November 1856, he left for the Granite City.

CHAPTER ELEVEN

NO JOKES ARE UNDERSTOOD HERE

1856–1860

In the space of a few months, twenty-five-year-old Maxwell had acquired two weighty responsibilities, and he was determined to discharge both of them to the best of his ability. He had taken care to make a good start as laird of Glenlair but no doubt felt a mixture of excitement and anxiety as he journeyed north, not being sure what awaited him in Aberdeen. It wasn't unusual for professors to be appointed young—Thomson had taken his chair in Glasgow at twenty-two, and Tait his in Belfast at twenty-three—but at this time Marischal College had a rather elderly collection: the youngest of Maxwell's colleagues was forty and their average age was fifty-five. The new professor must have seemed to them a mere boy, but they made him welcome—he was invited out so often that he rarely dined at his lodgings. They seemed to be pleased to have somebody young to talk to, but there was none of the free banter that Maxwell had expected. Aberdeen felt a long way from Cambridge. He wrote to Lewis Campbell:

> No jokes of any kind are understood here. I have not made a joke for 2 months, and if I feel one coming on I shall bite my tongue.[1]

Education was, indeed, serious business and Maxwell buckled down to it alongside his rather dour colleagues. Like other Scottish universities, Marischal College aimed to provide a broad and accessible education. Its staple four-year MA course was very broad indeed, taking in Greek, Latin, mathematics, natural philosophy, natural history, moral philosophy, and logic. Like the other professors, Maxwell had complete charge not only of the lectures but also of the syllabus and the exams. Avoiding any attempt at a joke, he set out his intentions in his inaugural address:

> My duty is to give you the requisite foundation and to allow your thoughts to arrange themselves freely. It is best that every man be settled in his own mind, and not be led into other men's ways of thinking under the pretence of studying science. By a careful and diligent study of natural laws I trust that we shall at least escape the dangers of vague and desultory modes of thought and acquire a habit of healthy and vigorous thinking which will enable us to recognise error in all the popular forms in which it appears and hold fast truth whether it be old or new.[2]

And, to be able to think for themselves, the students would need to *see* for themselves by doing experiments:

> I have no reason to believe that the human intellect is able to weave a system of physics out of its own resources without experimental labor. Whenever the attempt had been made it has resulted in an unnatural and self-contradictory mass of rubbish.

He carefully planned his course of lectures and agreed to give extra weekly evening classes at the Mechanics' Institution. We'll see shortly how well he succeeded.

In February 1857, he decided to send a copy of his paper "On Faraday's Lines of Force" to the great man. No doubt, he did so with some trepidation. After having read the *Experimental Researches in Electricity*, he felt tremendous empathy for Faraday, but he couldn't be sure that the feeling would be returned. He was, after all, an electrical novice—in the paper he had admitted that electricity was "a science in which I have hardly made a single experiment"—yet he had marched boldly into territory that Faraday had spent much of his life studying.[3] He needn't have worried. As we've seen, Faraday's response was grateful, gracious, and charming. The two had at once formed a rare bond.

A sign of the respect and trust that Faraday immediately felt for his young colleague was that he sent a paper of his own, asking for an opinion. It was the one that tentatively proposed *gravitational* lines of force, an idea generally thought outrageous. Faraday was aware that he was exposing himself to possible criticism by sending perhaps his most speculative paper, but he need not have worried. Busy with other work, Maxwell took his time replying but then offered a thought that must have surprised even Faraday—the idea could work if the lines of force were not attractive but *repulsive*. They would emanate from all the matter in the universe and two bodies in relative proximity, such as Earth and the sun, would be, as it were, in each other's shadow, and so would be *pushed* together. So they would appear to be attracted to one another. Moreover, the force of apparent attraction would follow an inverse-square law, and so would be indistinguishable from Newton's law of direct gravitational attraction. Faraday was impressed and grateful, but he replied apologizing for his own presumption in pressing Maxwell for a view.

> It was very wrong; for I do not think any one should be called upon for the expression of their thoughts before they are prepared, and wish to give them. I have often enough to decline giving an opinion because my mind is not ready to come to a conclusion.[4]

Questions of professional etiquette aside, it is clear that any reserve arising from differences in age and status had been swept aside.

Maxwell took up his work on colors again, but most of his spare time in the first few months in Aberdeen was taken up by Saturn's rings. The planet Saturn with its bizarre-seeming flat rings was the most mysterious object in the night sky and St. John's College, Cambridge, had chosen it as the theme for their prestigious Adams Prize, a biennial competition that had been founded to commemorate John Couch Adams's discovery of the planet Neptune. They had asked: under what conditions (if any) would the rings be stable if they were (1) solid, (2) fluid, or (3) composed of many separate pieces of matter? It was a fearsomely difficult problem that had defeated many mathematical astronomers, including the great Pierre Simon Laplace—it seemed the examiners had set it more in hope than in expectation.

Maxwell was, like everyone, intrigued by the rings, and to him they seemed to pose a giant, real-life version of the kind of problem he had tackled in the Smith's Prize examination. By a triumph of determination as much as skill, he managed to prove that solid rings would inevitably break up, and that the same would happen to fluid ones as the tidal waves grew bigger and bigger. He had thus shown that although the rings appear to be continuous, they must consist of many separate bodies orbiting independently—exactly the structure we have now seen on fly-past pictures from the Voyager and Cassini space probes. He tidied up his mass of calculations, posted off an essay weighing 12 ounces, and hoped for the best. It turned out that his was the only entry: the problem was so difficult that no one else got far enough to make it worth sending one in. He was awarded the Adams Prize, and the name James Clerk Maxwell began to be mentioned in high academic circles. The Astronomer Royal, Sir George Airy, described Maxwell's essay as "one of the most remarkable applications of mathematics to physics that I have ever seen."* It was, indeed, a triumph. All the mathematical methods that Maxwell used had been known for years—what was new was the combination of boldness, imagination, ingenuity, and sheer persistence that he had brought to the task.

Maxwell's jokes were, for the present, confined to letters, and Thomson, by now well established as professor of natural philosophy at Glasgow University, was one of his many correspondents. Thomson was now also a technical consultant on the Atlantic telegraph-cable project and came in for some gentle ribbing from Maxwell when the cable-laying ran into difficulties. On a visit to Glasgow by train, Maxwell wrote "The Song of the Atlantic Telegraph Company"—prompted, it seems, by the clickety-clack rhythm of the train wheels going over joints in the track. One of its verses runs:

Under the sea, under the sea,
No little signals are coming to me
Under the sea, under the sea,
Something has surely gone wrong,
And it's broke, broke, broke;
What is the cause of it does not transpire
But something has broken the telegraph wire
With a stroke, stroke, stroke,
Or else they've been pulling too strong.[5]

There wasn't an ounce of schadenfreude here; Maxwell was a great admirer of the project and thought the cable layers were doing a heroic job. He just couldn't resist having a little fun at their expense.

Romance was probably the last thing Maxwell expected to find in Aberdeen, but it happened. His partner in love was Katherine, daughter of the college principal, the Reverend Daniel Dewar. An invitation to dinner at the house led to many more, and the Dewars began to treat him as one of the family, even asking him to join them on a vacation. There, he proposed and Katherine accepted. It was an unusual match. Katherine Dewar was seven years his senior and appears to have taken little part in the intellectual life he enjoyed with his friends—we can even wonder whether she shared in his jokes. She has come in for harsh judgment from most writers, who report that she could be difficult and even jealous at times, but some of these impressions have originated from people who didn't know her very well, or who had their own axes to grind. As we shall see, there are points in her favor and points against. What is certain is that both had been lonely and shared the joy, and perhaps relief, in having found a lifetime companion. As was his way, Maxwell expressed his feelings in verse:

Trust me spring is very near
All the buds are swelling
All the glory of the year
In those buds is dwelling

What the open buds reveal
Tells us—life is flowing;
What the buds, still shut, conceal,
We shall end in knowing.

Long I lingered in the bud,
Doubting of the season
Winter's cold had chilled my blood—
I was ripe for treason.

Now no more I doubt or wait
All my fears are vanished,
Summer's coming dear, though late,
Fogs and frosts are banished.[6]

It was truly a joyful time, as is evident from the way metaphors fly even more exuberantly than usual in his letters. He had recently been best man at the wedding of Lewis Campbell, who was now an Anglican parish priest in Brighton, and naturally wanted his closest friend to do the same for him. Here, he tells Campbell of the engagement and the approximate date of the wedding, and invites him to bring his wife to visit Glenlair:

> We had done with the eclipse today, the next calculation was about the conjunction. The rough approximations bring it out early in June. . . . The first part of May I shall be busy at home. The second part I may go to Cambridge, to London, to Brighton, as may be devised. After which we concentrate ourselves at Aberdeen by way of concerted tactics. This done, we steal a march, and throw our forces into the happy valley, which we shall occupy without fear, and we only await your signals to be ready to welcome reinforcements from Brighton.

After a month's honeymoon of "sun, wind and streams" at Glenlair, Maxwell got back to work. One might be forgiven for thinking that he had abandoned electricity and magnetism, but the subject was never far from his mind and ideas were, as he put it, "fermenting and decocting." Meanwhile, on quite a different topic, he made a discovery that was truly a stroke of genius. Had he done nothing else, it alone would have been enough to set his mark on scientific history. His thoughts were set racing when he picked up a paper by the German physicist Rudolf Clausius on the rate of diffusion in gases—exemplified by the time it takes for the smell of perfume to cross a room when a bottle is opened. Clausius was a follower of the kinetic theory of gases, originally proposed by the Swiss physicist and mathematician Daniel Bernoulli, which attributed properties like temperature and pressure to the motion of molecules in the gas. Atmospheric pressure, for example, could be explained this way, but only if the air molecules moved very fast—hundreds of meters per second. Why, then, did the smell of perfume take several seconds to cross a room? Clausius had an explanation that was convincing yet mind-boggling. The molecules are forever colliding and changing direction—by the time it crossed a room, a single molecule would actually have traveled several kilometers. The astounding rapidity of molecular movements was becoming apparent for the first time. Maxwell described it this way:

> If you go at 17 miles a minute and take a totally new course 17,000,000,000 times in a second, where will you be in an hour?[7]

Kinetic theory was becoming plausible and had won converts, even though no one could be sure that molecules even existed. But there was a sticking point. Temperature was thought to depend on the speed of the molecules—the faster, the hotter—but, at a given temperature, did all the molecules travel at the same speed? This seemed unlikely, but, if not, what was the distribution of speeds, and how on earth could you work it out? Maxwell solved the problem in a few short paragraphs in what seemed almost like a conjuring trick and showed the distribution to have the shape of a lopsided bell. This was the Maxwell distribution of molecular speeds—the first ever statistical law in physics.[8] He had opened the door to great new regions of scientific knowledge—in particular to a proper understanding of thermodynamics, to statistical mechanics, and to the use of probability distributions in quantum mechanics.

Maxwell wrote up his new law and in the same paper made the important and surprising prediction that the viscosity of a gas, its internal friction, was independent of pressure. This happened because, at higher pressure, the dragging effect on a moving body of being surrounded by more molecules was exactly counteracted by the screening effect they provided. It was vital for the prediction to be tested by experiment—a verdict of false would demolish the whole kinetic theory, but a verdict of true would greatly strengthen it. As we will see, Maxwell later managed to do the experiment at home, with much help from Katherine. Elsewhere in the paper Maxwell made mistakes; he was off by a factor of 8,000 in one calculation because he had forgotten to convert kilograms to pounds and hours to seconds! Despite the flaws, his paper "Illustrations of the Dynamical Theory of Gases" drew gasps of admiration and put Maxwell in the first rank of physicists. However, the first person to recognize Maxwell's full achievement in bringing statistics into physics was at that time a schoolboy in Vienna, and he didn't see the paper until five years later. Ludwig Boltzmann was then so inspired by Maxwell's work on kinetic theory that he spent most of his career developing the subject further. The two began a kind of tennis match that lasted all of Maxwell's life; each in turn would be inspired by the other's work and counter with a further extension of the theory. Though they never thought of themselves as such, they were, in effect, a magnificent partnership, and it is pleasing that their names are linked in the Maxwell-Boltzmann distribution of molecular energies.

What of Maxwell's classes? For all his strong and progressive ideas on teaching, he was, sadly, not very good at it himself. Yet the students liked him. They were allowed to borrow only two books at a time from the college library, but Maxwell took out more for them, something professors were allowed to do for friends, and, when challenged, he replied that the students *were* his friends. He prepared his lessons carefully and would start well but then be drawn into what Lewis Campbell called "the spirit of indirectness and paradox that, though he was aware of the dangers, would often take possession of

André Marie Ampère. (Used with permission from Ken Welsh/The Bridgeman Art Library.)

Humphry Davy. (Used with permission from the Royal Institution, London, UK/The Bridgeman Art Library.)

The Royal Institution of Great Britain in the 1830s. Watercolor by Thomas Hosmer Shepherd. (Used with permission from the Royal Institution, London, UK/The Bridgeman Art Library.)

Faraday's induction ring, as it appears today. (Used with permission from the Royal Institution, London, UK/The Bridgeman Art Library.)

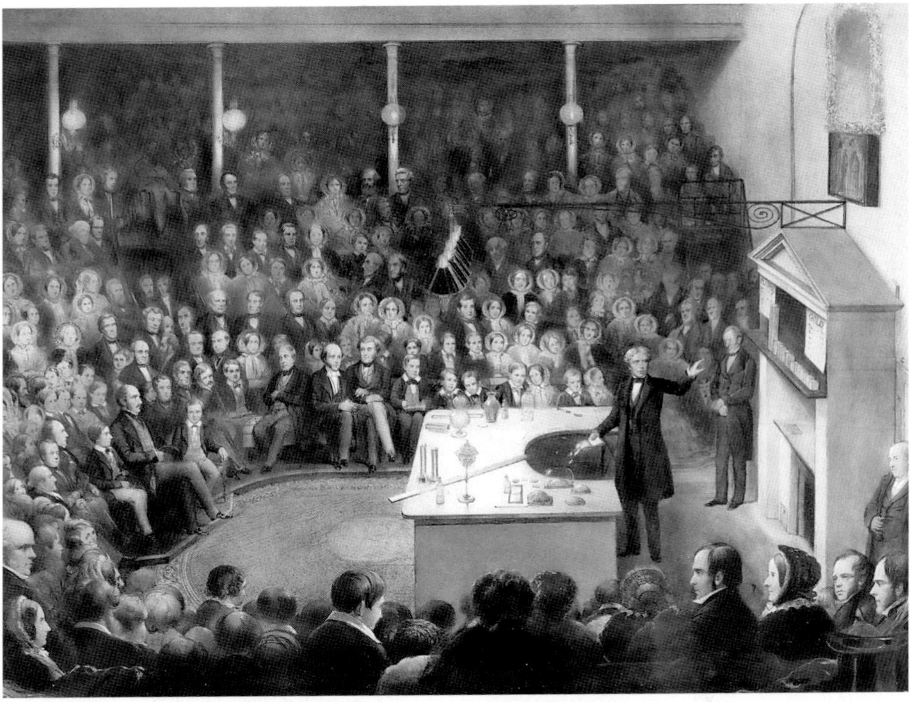

Faraday lecturing in the Royal Institution Lecture Theatre. (Used with permission from the Royal Institution, London, UK/The Bridgeman Art Library.)

Faraday in his laboratory. Engraving after a painting by Harriet Jane Moore. (Used with permission from the Royal Institution, London, UK/The Bridgeman Art Library.)

Portrait of Michael Faraday holding a bar magnet. (Used with permission from the Royal Institution, London, UK/The Bridgeman Art Library.)

Maxwell in his early thirties.
(Courtesy of the Master and Fellows of Trinity College, Cambridge.)

Above: Maxwell, aged twenty-four, holding his color top. (Courtesy of the Master and Fellows of Trinity College, Cambridge.)

Right: James Clerk Maxwell in his mid-forties. (Engraving by G. J. Stodart from a photograph by Fergus of Greenock. From Edinburgh and Scottish Collection, Edinburgh City Libraries.)

Left: Katherine Clerk Maxwell. (Courtesy of the Master and Fellows of Trinity College, Cambridge.)

Below: William Thomson (Lord Kelvin). (Used with permission from Universal History Archive/UIG/The Bridgeman Art Library.)

Above: Oliver Heaviside. (Used with permission from the Institute of Engineering and Technology, Hertfordshire, UK.)

Left: Heinrich Hertz during his military service. (Used with permission from the Institute of Engineering and Technology, Hertfordshire, UK.)

him against his will." He would throw in illustrations and metaphors that were intended to help but left most of the class bewildered. To add to his troubles, he made many algebraic errors on the blackboard that took time to find and correct. Many students, nevertheless, remembered him with affection. One reports:

> But much more notable [than the other professors] was Clerk Maxwell, a rare scholar and scientist as the world came to know afterwards; a noble-souled Christian gentleman with a refined delicacy of character that bound his class to him with a devotion which his remarkably meagre qualities as a teacher could not undo.[9]

And, to some, he was truly inspiring. David Gill, who became director of the Royal Observatory at the Cape of Good Hope, recalled:

> After the lectures, Clerk Maxwell used to remain in the lecture room for hours, with three or four of us who desired to ask questions or discuss any points suggested by himself or ourselves, and would show us models of apparatus he had contrived and was experimenting with at the time, such as his precessional top, color box, etc. These were hours of the purest delight to me.[10]

Gill had less fond memories of Katherine. The delightful afternoon sessions sometimes ended, he said, when Maxwell's "awful wife" appeared and called him home to an early dinner. Perhaps the early dinners were on the days when Maxwell gave his evening classes at the Mechanics' Institution. When talking to the working men, he seemed to be able to avoid the "spirit of indirectness and paradox," and his classes were remembered long after he left Aberdeen. One farmer recalled how his friend had been made to stand on a mat while the professor "pumpit him fu' o' electricity" so that his hair stood on end.[11]

It seems paradoxical that such a fine writer, who had strong and sound views on the principles of education, should have such travails in the classroom. As Campbell observed, Maxwell found it hard to bridge the gulf between his vast erudition and the students' modest compass. His quicksilver mind was always making connections, allusions, analogies, and comparisons that were quite beyond most students, but after all the years of free conversation with his father, who always understood his meaning from the slightest nuance, he found it hard to suppress them. The problem didn't arise on formal occasions, when he was obliged to slow down and present the words as he would when writing. We can be sure that in teaching, as in everything else, Maxwell did his best.

Marischal College was not the only university in Aberdeen; there was also King's College—this at a time when there were only three other universities in the whole of Scotland. Should they not unite, to gain economies of scale? Some important people thought so, and a royal commission had been set up to make a judgment on the matter. There was talk of "union"—common management of otherwise little-changed functions—but the commission decided in favor of "fusion," a complete merger that would halve the number of professors. The new, combined, University of Aberdeen needed only one professor of natural philosophy, and the man they chose was Maxwell's opposite number at King's. One reason for this extraordinary decision was that to lose Maxwell was the cheaper option, as he had not served long enough to qualify for a pension. Another was that his rival was well dug-in and a polished negotiator, who was known as "Crafty" Thomson. As for Maxwell's research, very few people at the time had any idea of its importance, and none of them lived in Aberdeen.

Just at this time, Maxwell heard that his old mentor James Forbes had accepted the principalship at St. Andrews University, leaving the chair of natural philosophy at Edinburgh vacant. This would be a wonderful job and, naturally, Maxwell applied. But the post was equally attractive to his old friend P. G. Tait, who wanted to return to Scotland from Belfast, and he applied, too. Once again they were rival candidates, and this time Tait was preferred. The electors' choice was not as strange as it may seem to us—Tait was a first-rate physicist and a fine lecturer with a commanding presence. He was also the first person Maxwell turned to for mathematical advice, when needed. Twice spurned in his own country, Maxwell looked elsewhere and saw that King's College, London, wanted a professor of natural philosophy. He applied for the post and was selected.

There was plenty to do meanwhile. Besides preparing his great paper on kinetic theory for publication, he wrote another, on elastic spheres, and sent a report on his color vision experiments to the Royal Society of London—work for which he was soon rewarded with the Society's Rumford Medal. At home there was estate business to see to and, as laird of Glenlair, he was expected to play a leading part in local affairs. Following his father's example, Maxwell took on this role wholeheartedly—for example, helping to raise funds for the endowment of a new church in the nearby village of Corsock. During the summer, he went to a horse fair and bought a handsome bay pony for Katherine. Soon after returning he became severely ill with a high fever. It was smallpox, almost certainly contracted at the fair, and it almost killed him. Maxwell was in no doubt that Katherine's devoted nursing had saved his life. He was laid up for several weeks, but strength and vitality came back little by little, and he was able to break in Charlie, the new pony, riding side-saddle with a carpet taking the place of a lady's riding habit. In October 1860, after an incident-packed year, he and Katherine made the long journey south to London.

CHAPTER TWELVE

THE SPEED OF LIGHT

1860–1863

King's College, in the Strand, had been founded in 1829 as an Anglican alternative to the new nonsectarian London University, now University College, which had itself been founded as a secular alternative to the strictly Church of England universities of Oxford and Cambridge. Its educational mission was to prepare young people for life and work in a rapidly changing world. Unlike the traditional fare provided by Cambridge and Aberdeen, its courses were much like those at today's universities. King's not only gave classes in modern subjects like chemistry, physics, botany, and economics but also ran purpose-built courses in law, medicine, and engineering.

At the age of twenty-nine, Maxwell delivered his second inaugural lecture. Experience had confirmed the soundness of the theme he had introduced at Aberdeen—his job was to help people think for themselves—and he developed it further:

> In this class, I hope you will learn not merely results, or formulae applicable to cases that may possibly occur in our practice afterwards, but the principles on which those formulae depend, and without which the formulae are mere mental rubbish.
>
> I know the tendency of the human mind is to do anything rather than think. But mental labor is not thought, and those who have with labor acquired the habit of application often find it much easier to get up a formula than to master a principle.[1]

He finished the lecture with a message that seemed to be addressed to himself as much as to the students and that turned out to be extraordinarily prophetic:

> Last of all we have the Electrical and Magnetic sciences, which treat of certain phenomena of attraction, heat light and chemical action, depending on conditions of matter, of which we have as yet only a partial and provisional knowledge. An immense mass of facts has been collected and these have been reduced to order, and expressed as a number of experimental laws, but the form in which these laws are ultimately to appear as deduced from general principles is as yet uncertain. The present generation has no right to complain of the great discoveries already made, as if they left no room for improvement. They have only given science a wider boundary, and we have not only to reduce to order the regions already conquered but to keep up operations on a continually increasing scale.

Within four years, he was to turn rhetoric into fact by opening up vast new regions of scientific knowledge.

The Maxwells rented a house in the newly developed district of Kensington, close to the big open space of Hyde Park and Kensington Gardens, which was a fine place for strolling and for riding. James had a vigorous four-mile walk to work on fine days, with the alternative of a horse-drawn bus ride. The walk took him first through the park and then along Piccadilly, passing within a few yards of the Royal Institution in Albemarle Street. Faraday was, by now, retired and living at Hampton Court, but he still called in regularly at the Institution and, though there is no record of it, we can be fairly certain that he and Maxwell met there from time to time for a chat. It was probably at one such occasion that Faraday asked Maxwell to give one of the Institution's now-famous Friday Evening Discourses. Maxwell naturally accepted and chose to talk about color vision.

It was a perfect occasion to demonstrate the three-color principle—the eye's three sets of receptors channel their separate signals to the brain, which then combines them to manufacture the color that you "see." But Maxwell's color top was far too small for people in the back seats to see clearly and his color box could only be used by one person at a time. Something else was needed—what about a color photograph? The techniques of black-and-white photography were by now well-known, and one of his new colleagues, Thomas Sutton, was an expert. They devised a simple scheme. Take three ordinary photographs of the same object, one through a green filter, one through a red filter and one through a blue filter; then project them through the same filters on to a screen, superimposing the three beams of light to form a single image. The experiment worked beautifully; the audience at the Royal Institution sat spellbound as the image of a tartan ribbon appeared on the screen in glorious color. Maxwell had produced the world's first color photograph.[2]

The time had come to give voice to the thoughts on electricity and magnetism that had been forming in "the department of the mind conducted independently of consciousness." In his first paper, six years earlier, he had taken the flow of a hypothetical weightless fluid as an analogy and shown that the known formulas for static electric and magnetic fields did not depend on the orthodox assumption that forces resulted from material bodies acting on one another at a distance; they could be derived equally well from Faraday's idea of lines of force in space. As we've seen, Maxwell was struck early on by the integrity and power of Faraday's writings, and the years of subconscious "decocting" of ideas had served to convince him more and more that Faraday was right—fields of force truly *existed* in space.

In his inaugural address, Maxwell had, in effect, set himself a manifesto: to produce a theory that explained all the known experimental laws of electricity and magnetism by deduction *from general principles*. These laws were, briefly:

1. Like electric charges repel one another and unlike ones attract, both with a force inversely proportional to the square of the distance separating them.
2 Like magnetic poles repel one another and unlike ones attract, both with a force inversely proportional to the square of the distance separating them. But poles exist only in north/south pairs and all magnetism, even in permanent iron magnets,

probably results from electric currents. (Law 3 implies that any loop of current acts as a magnet with a north pole on one side of the loop and a south pole on the other.)

3. A current in a wire creates a circular magnetic field around the wire, its direction depending on that of the current.
4. A changing magnetic field, or flux, that passes through a conducting circuit generates an electric current in the circuit, its direction depending on whether the flux is increasing or decreasing.[3]

No satisfactory complete theory existed, though Wilhelm Weber had made a brave and ingenious attempt. His highly mathematical theory, based on action at a distance, required the force between electric charges to depend not only on their distance apart but also on their relative velocity and acceleration. Much as he respected Weber's work, Maxwell's intuition bridled at these assumptions and at the whole concept of action at a distance. He felt sure that the true theory lay instead on the road indicated by Faraday and that his best chance of finding it was by a suitable analogy.

He sought a mechanical analogy that could represent changing fields as well as static ones. A tall order, but his thoughts eventually led to a promising idea. There were reasons to suppose that something *rotational* might be going on in a magnetic field. For one thing, this could help explain why magnetic force acted in a circle around an electric current. For another, Faraday had shown that when polarized light passed through a strong magnetic field, its plane of polarization of light was rotated. The epitome of rotation was a vortex in a fluid, and a vortex in a fluid had a natural tendency to contract along its spin axis and expand sideways. One could imagine all space filled with a fluid in which vortices could exist, and, in a region of space, a system of adjacent vortices that were all rotating the same way with their axes parallel. There would be a tension along their lengths and each would exert a sideways pressure against its neighbors. This property was exactly analogous to Faraday's magnetic lines of force, which exerted a tension along their lengths and a repulsive pressure against each other.

As his thoughts developed, Maxwell came to replace the fluid vortices in his model with solid, tiny, close-packed spherical cells that could spin. As it spun, each cell would tend to expand at its equator and flatten at its poles, so the combined effect of many cells spinning with their axes aligned would be exactly the same as the vortices—longitudinal tension and sideways pressure, again corresponding to the properties of Faraday's magnetic lines of force. For simplicity, we'll make the change from vortices to cells now, a little earlier than Maxwell did.

Two more components were needed: something to set the cells spinning and something to prevent the edges of neighboring cells from rubbing awkwardly against each other. Maxwell solved both problems in a single stroke. To stop neighboring cells from rubbing against each other, he put even smaller particles between them to act like ball bearings or like the "idle wheels" that an engineer places between two gears that need to rotate in the same direction. Then came the inspiration: suppose these tiny particles were *particles of electricity*. In the presence of an electromotive force, they would move along channels between the cells, constituting an electric current, and it was this movement that set the cells spinning.

The contraction of the spinning cells along their aligned axes of spin represented magnetic lines of force; the faster the cells spun, the greater the contraction and the stronger the force. And the north-to-south direction of the forces, by Maxwell's convention, was the way a right-handed screw would move if it rotated the same way as the cells. The expansion of the spinning cells around their "equators" represented the sideways repulsion between, magnetic lines of force. Maxwell was on his way, but he still needed to accommodate another of Faraday's discoveries: different substances had different magnetic characteristics. Some, like iron and nickel, had a high magnetic inductive capacity (they conducted magnetic lines of force very well), while others, like wood, had a lower inductive capacity even than a vacuum (they were the diamagnetic materials). Maxwell solved the problem with his customary sureness of touch. His imaginary cells were everywhere, coexisting with any ordinary matter that occupied the same space, and he simply made the density of each cell proportional to the inductive capacity of whatever ordinary substance was present in the same space—the denser the cell, the more readily the substance conducted magnetic lines of force. By the same token, the higher the density of the cells, the greater would be the forces of longitudinal contraction and sideways expansion for a given rate of spin. In Maxwell's analogy, these forces represented the concentration, or density, of magnetic flux. But if the cells were everywhere, why were they not apparent, and how could they coexist with ordinary matter? Maxwell wasn't deterred by such awkward questions. The mass density of the cells could be very low indeed, so low that they offered no significant obstruction to ordinary matter and hence were undetectable by any known instrument. As long as they had *some* mass and rotated fast enough they would contract along their axes of spin and so generate the necessary forces. And, in any case, it was only a conceptual model, an aid to thought.

Density wasn't the only property of the cells to vary in this way. In a part of space occupied by an insulator, the cells, or perhaps local groups of cells, would hang onto their particles of electricity; but in a good conductor, like a copper wire, the particles could move freely. This "stickiness" represented the electrical resistance of the material—an ideal conductor would have none, an ideal insulator would be perfectly sticky, and real-world materials filled the range in between. The tiny particles of electricity had rolling contact with the cells; there was no sliding. In a uniform, unchanging, linear magnetic field, the particles wouldn't move bodily; they would just rotate, along with the cells. But if a row of particles moved without rotating, thus forming an electric current, they would set the cells with which they had contact spinning—exactly the conditions needed to create a circular magnetic field around a current-carrying wire, law 3 above. If the particles rotated *and* moved, the circular field due to their movement would be superimposed on the linear one due to their rotation. In law 2, the magnetic forces had already been explained, and the inverse-square rule was intrinsic to the model—although the mechanism was more complicated than that in Maxwell's fluid model of his Cambridge days, this was, once again, essentially a matter of geometry.[4]

Next to be conquered by Maxwell's model was law 4: a changing magnetic field that passes through a conducting circuit generates an electric current in the circuit—Faraday's law of induction. Maxwell chose to demonstrate an equivalent effect—that when a current is switched on in one circuit, it induces a pulse of current in a nearby but separate circuit by creating a changing magnetic field that links the two. This was exactly the effect that Faraday had discovered in his iron-ring experiment of 1831, and Maxwell explained in detail how his model simulated it. He drew a diagram, giving the cells a hexagonal cross section "purely for artistic reasons," and we can see it, slightly adapted for our purpose, in Figures 12.1a–d.

The diagrams show a cross section of a small region of space. The idle-wheel particles along AB are in a wire that is part of a circuit with a battery and a switch, initially open. Those along PQ are in another wire that is part of a separate circuit having no battery or switch. The idle wheels along AB and PQ are free to move because they are in conductors, but others in the neighborhood are in a nonconducting material and can only rotate in their fixed positions. AB and PQ are, of course, impossibly thin wires and impossibly close together, but this is to just keep the diagram compact; the argument Maxwell produced would apply equally well to normal-sized and normally spaced wires containing many rows of idle wheels and cells. The argument runs like this:

Suppose the magnetic field is zero at first, and the switch open, so that all the cells and idle wheels are stationary (figure 12.1a). When the battery is brought into the circuit by closing the switch, the idle wheels along AB move bodily from left to right without rotating, constituting a current. This causes the rows of cells on either side of AB to rotate in opposite directions, thus creating a circular magnetic field around the wire. The idle wheels in PQ are now pinched between rotating cells on the AB side and stationary ones on the other, so they start to rotate (clockwise) and also to move from right to left, the opposite direction from those in AB (figure 12.1b).

But the circuit containing the wire along PQ has some resistance (all circuits do), so the idle wheels there, after their initial surge, will slow down, causing the cells above PQ to begin rotating counterclockwise. Soon, the sideways movement of the idle wheels will stop, although they will continue to rotate. By this time, the cells above PQ will be rotating at the same rate as those in the row below PQ (figure 12.1c).

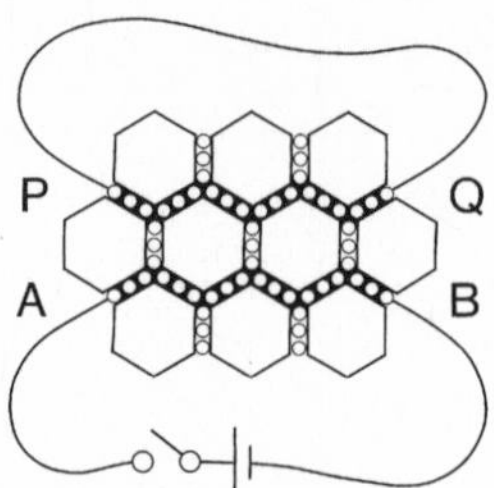

Fig. 12.1a. Switch open:

- All cells and idle wheels are stationary.
- There are no currents.
- There is no magnetic field.

(Used with permission from John Bilsland.)

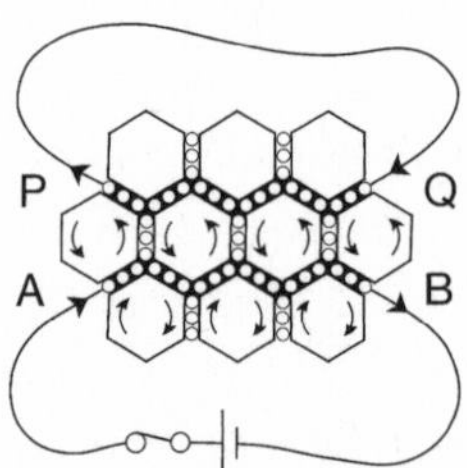

Fig. 12.1b. Switch first closed:

- AB current flows from left to right.
- PQ current flows from right to left.
- Cells below AB rotate clockwise, causing a magnetic field pointing away from the viewer.
- Cells between AB and PQ rotate counterclockwise, causing a magnetic field pointing toward the viewer (in three dimensions, a circular field envelopes AB).
- Cells above AB are still stationary.

(Used with permission from John Bilsland.)

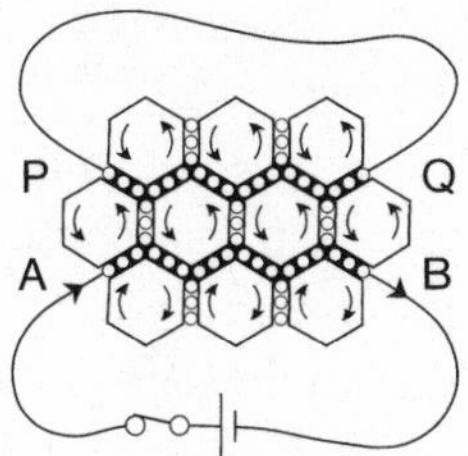

Fig. 12.1c. Shortly after switch closed:

- PQ current slows, then stops.
- Cells above AB start to rotate counterclockwise and by the time the current stops are rotating at the same rate as those in the row below PQ.

(Used with permission from John Bilsland.)

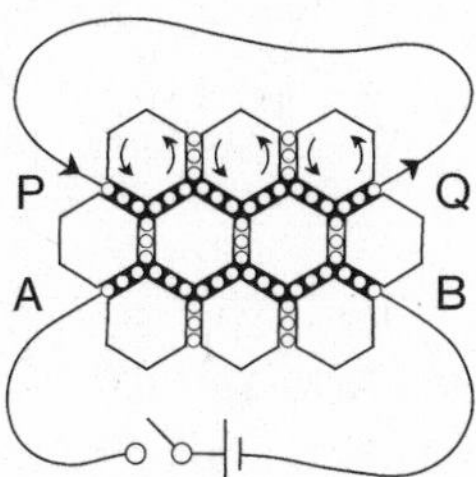

Fig. 12.1d. Switch opened again:

- AB current stops.
- Cells in rows above and below AB stop rotating.
- PQ current flows from left to right.
- The current will slow, then stop; the situation will then be as in figure 12.1a.

(Used with permission from John Bilsland.)

When the switch is opened again, disconnecting the battery, the idle wheels along AB stop moving and the rows of cells on either side of AB stop rotating. The idle wheels in PQ are now pinched between stationary cells on the AB side and rotating ones on the other, so they start to move from left to right, the same direction as the original AB current (figure 12.1d). Once again the resistance of the circuit containing PQ causes the idle wheels there to slow down. This time, when their sideways movement stops, they will not be rotating; we are back to the state represented in figure 12.1a.

Thus switching on a steady current in AB induces a pulse of current in PQ in the opposite direction, and switching the current off induces another pulse in PQ, this time in the same direction as the original current. More generally, any *change* of current in the AB circuit induces a current in the separate PQ circuit through the changing magnetic field that links them. Equivalently, any change in the amount of magnetic flux passing through a loop of wire induces a current in the loop; law 4 is explained. If the battery in the AB circuit were replaced by an alternating current generator, the alternating AB current would induce an alternating current in PQ. This is exactly the way transformers work in our electrical power supply systems.

And here, at last, was a physical interpretation of Faraday's electrotonic state. Faraday had thought this state to be some kind of strain that was present in a wire whenever the wire was immersed in a magnetic field but that showed itself only when the field changed. In his iron-ring experiment, for example, a brief current had appeared in his secondary circuit when the magnetic field around it collapsed because the primary current had been switched off. Maxwell's interpretation was different, though the effect was the same. The cells in his model had inertia and so acted as a store of rotational momentum when they spun. Any change in this momentum was accompanied by a force, rather like the force that throws you forward out of your seat when a bus stops suddenly, and this took the form of an electromotive force that would drive a current along any conducting path—in his model it drove a row of the tiny idle wheels that represented particles of electricity. Faraday's electrotonic state was a sign of what

Maxwell called "electromagnetic momentum," and it had a defined value at every point in the field.[5]

The most elusive law turned out to be law 1, on the forces between electric charges, commonly called "electrostatic forces," and for the present Maxwell saw no way of bringing them into the model. It was disappointing not to have achieved a full theory, but he wrote up his results with full mathematical rigor and in the spring of 1862 published them in two parts in a paper called "On Physical Lines of Force."[6] As he had done when presenting his earlier paper "On Faraday's Lines of Force," he took care to warn readers not to take the model literally:

> I do not bring it forward as a mode of connexion existing in nature, or even as that which I would willingly assent to as an electrical hypothesis. It is, however, a mode of connexion which is mechanically conceivable, and easily investigated, and it serves to bring out the actual mechanical connexions between the known electromagnetic phenomena; so that I venture to say that any one who understands the provisional and temporary nature of this hypothesis, will find himself rather helped than hindered by it in his search after the true interpretation of the phenomena.[7]

There, it seemed, things would stay, but during the summer holiday at Glenlair, an idea began to crystallize. To transmit internal forces across their own bodies without losing energy, his cells would need to have a degree of springiness, or elasticity. Could this explain the forces between electric charges? In insulators, the little "idle wheels" that represented particles of electricity couldn't move freely, as they could in conductors—they were held captive, as it were, by their parent cells. But, when an electromotive force tried to move the particles in an insulator, adjacent *elastic* cells would distort, allowing the particles to move a short distance. The distorted cells would try to spring back to their original shape, exerting a restoring force—the greater the distortion, the greater the force—and the particles would move until this force was sufficient to balance the electromotive force. The short movement of the little electrical particles represented a general displacement of electricity within the insulating material. If, as seemed likely, all matter was composed of molecules, the displacement of electricity would occur within each molecule; in other words, the molecules would become electrically polarized, just as Faraday had surmised. (Although Faraday was not convinced of the existence of atoms, he did believe that small particles of matter became polarized in this way.) But Maxwell was about to enter completely new territory. His assembly of cells and idle wheels pervaded all space, whether or not the space was occupied by ordinary matter. So, according to the model, the electrical displacement would take place even in a vacuum where there were no molecules to be polarized!

Faraday had found that substances varied in their ability to conduct electric lines of force, formally their electric inductive capacity, and Maxwell accommodated this in his model by endowing each cell with a degree of elasticity that corresponded to the electric inductive capacity of whatever ordinary material occupied the same space. If this material was an insulator subjected to an electromotive force, the distance moved by the model's little electrical particles would depend not only on the strength of the force but also on the inductive capacity of the material. If the material had a high inductive capacity, the cells had a soft springiness, so enabling the particles to move a relatively long way; but if it had a very low inductive capacity, the cells were stiff, scarcely letting the particles move at all. The distance moved by the particles represented the electrical displacement in the material, and the displacement constituted electrical induction, or flux. This matched Faraday's finding exactly: for a given electromotive force, the electric flux in a material was proportional to its inductive capacity.

The electrostatic forces in law 1 were explained, along with the inverse-square rule, which, as for magnetic forces, was intrinsic to the model and essentially a matter of geometry.[8] The explanation of the forces matched Faraday's idea that they were a manifestation of some kind of strain in the insulating material. In the model, the strain was in the distorted cells, each trying to regain its original shape. To Maxwell, as it would have been to Faraday, this explanation was physically more satisfying than the mysterious action at a distance favored by most people. For example, the electrostatic force of attraction between two oppositely charged bodies wasn't the result of the bodies somehow attracting one another at a distance along the straight line between them. It arose because the strain in the surrounding insulating material (which could be just empty space) acted on *both* bodies, pushing them together.

Maxwell had shown how the electrical and magnetic forces that we experience could have their seat not only in material objects like magnets and wires but in a form of energy that existed in the space that surrounded them. Magnetic energy was akin to kinetic energy, the energy of mass in motion, like that of a moving train or the flywheel on an exercise bicycle, and electrical energy resembled potential energy, like that in a wound-up spring. And the two forms of energy were inextricably linked—a change in one was always accompanied by a change in the other. He had demonstrated how they acted together in accordance with the laws governing all known electrical and magnetic phenomena. A stupendous achievement, but things didn't stop there: the model predicted two new phenomena, both so remarkable that no one could have foreseen them.

Maxwell made the extraordinary claim that brief electric currents could exist in a material that was a perfect insulator. It was a simple consequence of his notion of electrical displacement: the small movements of the tiny particles between the cells that occurred *during the act of displacement* were, in effect, brief electric currents. Maxwell called them "displacement currents." Moreover, his cells and particles pervaded every region of space, whether or not that region was shared with ordinary matter. So, according to the model, electromotive forces would act on a vacuum in exactly the same way that they did on any other insulating material: the particles would move a short distance while their parent cells distorted. In other words, displacement currents would occur even in empty space! He had found the elusive final link that united electricity and magnetism. The known laws of electricity and magnetism had lacked symmetry and completeness, but with the displacement current everything fit together in a compact and beautiful theory. This wasn't immediately evident, however, even to Maxwell; he had seen something else.

Any medium that possessed both elasticity and inertia should be able to transmit *waves*, and his had both. He considered how fluctuations in the electric and magnetic fields would spread through the assembly of cells. As we've seen, whenever an electromotive force was first applied to an insulating medium, even empty space, there would be a brief electric current because the little particles would move a short distance before being halted by the spring-back force of their parent cells. This movement would then be transmitted through the cells on each side to neighboring rows of particles, and then on through the next layer of cells to particles beyond. The process wouldn't be instantaneous, because internal elastic forces had to be transmitted *through* each cell and they needed time to overcome the inertia of the cell's mass. So every twitch of the particles, accompanied by a shimmy of the cells, would spread out in a wave motion. The twitches of the particles represented fluctuations of the electric field, and the twisting movements of the cells represented those of the magnetic field. The two were inseparable and they combined to send out waves of energy through all space. Maxwell had predicted *electromagnetic waves*. Faraday's "shadow of a speculation" in his 1845 "Ray-vibrations" lecture—that oscillations in lines of electric and magnetic lines of force would propagate as waves—now stood on firmer, though still controversial, theoretical ground.

Mathematical scientists had studied wave motion and identified two known types. Those in which local movement was parallel to the line of wave travel, like sound waves, were called *longitudinal*, or *compressional*, and those like waves in the sea or in a rope, where the movement was perpendicular to the direction of the waves, were called *transverse*. Remarkably, Maxwell's electromagnetic waves were *doubly* transverse: fluctuations in the electric and magnetic fields were at right angles to each other, and the wave direction was at right angles to both of them.

Light waves were known to be transverse, and this raised a compelling question: could they be a form of electromagnetic energy? The speed of light had been measured by experiment, and Maxwell had worked out that the speed of his waves in empty space, or air, was equal to the ratio of the electromagnetic and electrostatic units of electric charge.[9] This was a fundamental quantity that could only be determined by an exceedingly difficult experiment, and the experiment had been carried out only once—by Wilhelm Weber and his colleague Rudolf Kohlrausch. Maxwell would need to convert their result to a different system of units, but that was a simple matter. However, he had brought no reference books home to Glenlair, and the rest of the summer vacation passed in a glow of anticipation. Back at King's College in October, he looked up Weber and Kohlrausch's result, carried out his calculation, and made the comparison. The speed in a vacuum (or in air) of his predicted waves was 310,740 kilometers per second. Armand-Hippolyte-Louis Fizeau had measured the speed of light at 314,850 kilometers per second. The results were too close for coincidence—the difference of a little over one percent was well within the range of possible errors in the two experiments. Light must be electromagnetic.

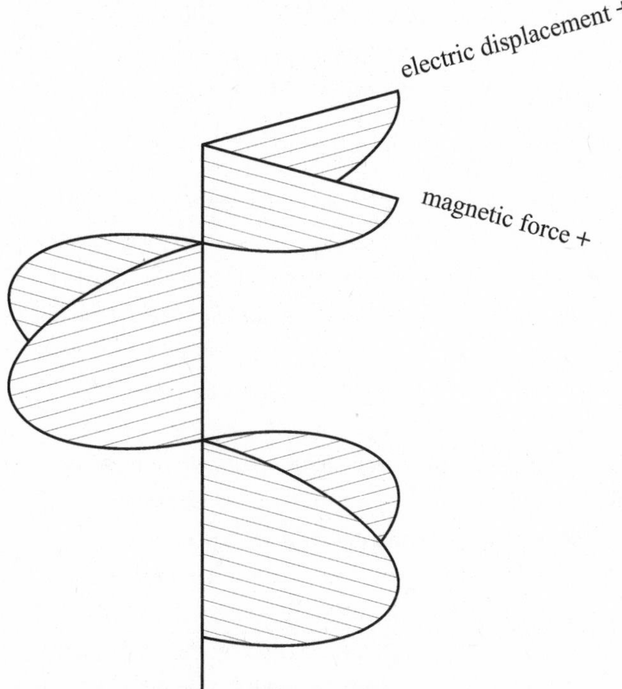

Fig. 12.2. Maxwell's illustration of an electromagnetic wave. (Used with permission from Lee Bartrop.)

He hadn't expected to extend his paper "On Physical Lines of Force" beyond parts 1 and 2, but now he set about writing part 3, on electrostatic forces, displacement currents, and waves, and part 4, which dealt with how the plane of polarization of polarized light was rotated by a magnetic field—the effect Faraday had discovered in 1845. In part 3, published early in 1862, Maxwell announced:

> We can scarcely avoid the inference that light consists in the transverse undulations of the same medium which is the cause of electrical and magnetic phenomena.[10]

He had united electricity, magnetism, and light—a stupendous achievement. Yet his announcement caused barely a ripple. As physicists generally believed that an aether of some kind was necessary for the propagation of light, one might have expected them to accept Maxwell's extension of the principle to electricity and magnetism. But his model seemed so weird and cumbersome that nobody thought it could possibly represent reality. The reaction of his friend Cecil Monro was typical:

> The coincidence between the observed velocity of light and your calculated velocity of a transverse vibration in your medium seems a brilliant result. But I must say I think a few such results are needed before you can get people to think that every time an electric current is produced a little file of particles is squeezed along between two rows of wheels.[11]

An obstacle was embedded deep in the scientific thinking of the time. Immersed in Newton's clockwork universe, people had thought that all physical phenomena resulted from some kind of mechanical action (combined, where appropriate, with forces such as gravity, which acted instantaneously at a distance) and that all would be clear to us if, and only if, we discovered the *true* mechanisms. Despite all his warnings, people couldn't understand that Maxwell's model didn't purport to represent nature's actual mechanism, but that it was merely a temporary aid to thought, a means of arriving at the relevant mathematical relationships by using an analogy. His analogy happened to use spinning cells, but that was by the way; it was the mathematical relationships that were important. Maxwell had conceded that the model was "somewhat awkward," and to many of his contemporaries it was nothing more than an ingenious but flawed attempt to portray the true mechanism, for which the search would continue.

Probably not even Maxwell recognized the full measure of his achievement. Using only the familiar tools and materials of Newtonian mechanics, he had succeeded in building a bridge to unimagined new regions of scientific knowledge. The bridge was a bizarre and ungainly construction, but it served the purpose, and the wonder was that it was built at all; nobody but Maxwell had seen a need for it. What was it about Maxwell that set him apart from his contemporaries? Two characteristics stand out.

The first may seem paradoxical: he was in one sense a truer follower of Newton than most of his predecessors and contemporaries. As we've seen, the first mathematical laws of electricity and magnetism had been modeled on the law of gravity. According to Newton, the gravitational force between two masses was proportional to their product divided by the square of the distance between them. Simply replace masses with charges or pole strengths, and you had the basic laws of electricity and magnetism. But with the work of Coulomb, Ampère, Poisson, and others had come an assumption that the forces resulted from instantaneous action at a distance between the masses, poles, or charges. Newton himself had been careful not to make any such assumption—indeed, he had, as we've seen, described action at a distance as "so great an absurdity, that I believe no man, who has in philosophical matters a competent way of thinking, can ever fall into it."[12] But this warning had been forgotten, and throughout the early and middle 1800s, the only prominent physicists to challenge action at a distance openly were Faraday and Maxwell.

The second characteristic that set Maxwell apart from his peers is epitomized by his prediction of displacement currents in empty space, and consequently of electromagnetic waves. No hint of either had come from experimental results; nor were they prompted by logic.

However long is spent on the search for an explanation, one is forced back to a single word—*genius*.

Faraday was by this time sinking into senility and was not up to reading Maxwell's paper, but if he had been able to give an opinion, it probably wouldn't have been wholly favorable. He had liked Maxwell's first paper, in which lines of force were represented, in analogy, by the smooth flow of a fluid, but this one possessed some features he wouldn't have liked at all. To Faraday, lines of electric and magnetic force were fundamental, freestanding, self-sufficient, but in Maxwell's model they had been, in a sense, demoted—they were merely the effect of the motion of his tiny cells and even smaller particles. And for all Maxwell's warnings that his model wasn't intended to portray nature's actual mechanism, it did *seem* to hypothesize that nature in some way operated like an assembly of atoms—objects that Faraday thought fanciful; he had said of them "why assume the existence of that of which we are ignorant, which we cannot conceive, and for which there is no philosophical necessity?"[13] Although Faraday had been the inspiration for Maxwell's model, one suspects that he would not have recognized in it much of his own view of physical reality. In particular, he would have objected to its reliance on a medium—in his vision, lines of force transmitted their own vibrations without the need for any medium. One aspect of reality, however, was beyond Faraday's reach—mathematical relationships—and the whole purpose of Maxwell's construction-kit model had been to discover these. The aim had been achieved, and what Maxwell sought now was a way of freeing the theory from any arbitrary physical hypotheses.

For now, he dispatched thoughts on electromagnetism to "the department of the mind conducted independently of consciousness" and gave his attention to an urgently needed experiment on another topic. He had shown in his first paper on the kinetic theory of gases that the viscosity of a gas should be independent of pressure, but the prediction had yet to be tested. It was a make-or-break test for the kinetic theory. If the prediction turned out to be wrong, the theory would be demolished; but if the experimental results showed it to be true, the theory would be greatly boosted. Nothing at all similar had been done before, so the laboratory facilities at King's didn't serve the purpose, and he decided to try the experiment at home. Maxwell is generally regarded as a cerebral genius, as indeed he was. But he also enjoyed practical work, and countless hours in his improvised laboratory-cum-workshop at Glenlair had honed his skill. The gas viscosity problem posed a formidable challenge to any experimenter, but it needed to be tackled, and Maxwell rolled up his sleeves.

In his Kensington attic, with Katherine as assistant, he carried out one of the most spectacular home experiments in the history of science. A tripod taller than a man held a torsional pendulum enclosed in a huge glass case connected by a tube to a pump that was used to raise or lower the pressure of the air that, by virtue of its viscosity, damped the swings of the pendulum. First the pressure seals failed, then the glass case imploded with a great bang, but Maxwell persevered and eventually got a reliable set of readings that emphatically verified his prediction that viscosity was independent of pressure—a milestone in the development of the kinetic theory of gases.

There were many other calls on his time, and some were in the cause of technological progress. In the same way that Faraday had answered calls to work on nationally important projects like optical glass and lighthouses, Maxwell came to the aid of the telegraph industry. One problem in particular needed attention because it bedeviled the greatest technical enterprise of the day—the laying of a properly working telegraph cable under the Atlantic Ocean. The first Atlantic cable, laid in 1858, had failed after a few weeks, and subsequent examination of recovered sections had shown it to be of poor quality. William Thomson led a drive to bring in proper quality control in the manufacture and supply of cables, and the most pressing need was for a physical *standard* of electrical resistance so that cables supplied by the manufacturers could be properly tested against specification.

William Thomson had suggested an ingenious experiment for the purpose, and Maxwell led a team from the British Association for the Advancement of Science to carry it out at King's College. His colleagues were fellow Scots—Fleeming Jenkin, who had also attended Edinburgh Academy, and Balfour Stewart. The idea was to spin a copper-wire coil rapidly in Earth's magnetic field and thereby generate a current in the coil that would have its own magnetic field. This field would vary as the coil rotated but would act predominantly either to the east or to the west, depending on which way the coil was spun. A magnetic needle, delicately suspended at the center of the coil, would swing back and forth but eventually settle at a fixed angle to Earth's field. The wonder of Thomson's design was that the angle of the needle's deflection depended only on the resistance of the coil, along with known factors like the dimensions of the coil and its speed of rotation. Using the appropriate formula, the angle gave *an absolute measure* of the resistance of the coil, which could then be used to calibrate a conveniently transportable "standard" model resistance, one that could be easily reproduced. The copies could then be taken anywhere and used to measure the resistance of lengths of cable, or anything else.

Putting the elegant design into practice wasn't easy. For every reading, the coil had to be hand-cranked for about nine minutes, keeping the speed constant, and Jenkin made a special governing device for the purpose. Many runs had to be aborted when there was a mechanical fault or when iron ships passing on the nearby river Thames distorted Earth's magnetic field, but months of patient work was finally rewarded: the world had its first standard of electrical resistance. And it soon had telegraph communication across the Atlantic Ocean. With William Thomson on the board of directors, the Atlantic Telegraph Company laid a sound cable in 1866, and more soon followed.

A characteristic of Maxwell's work, indeed his life, was that he seemed to take everything in his stride—he was never hurried. Somehow, he and Katherine managed to go riding in the park most afternoons and, of course, they went on accumulating data on color vision, asking all new houseguests to have a go. They had installed the latest big color box near the window in an upstairs room, and people across the road were alarmed at first to see them peering into what looked like a coffin. Maxwell also found time to keep up with the scientific journals and to pass any useful information on to his students. A case in point was William Rankine's analysis of forces in structures like steel-girder bridges, and here Maxwell brought in a dramatic improvement of his own. He introduced so-called reciprocal diagrams. Lines that converged to a point in the real structure became polygons in the new diagrams, and this made it easy to work out the forces graphically without the need for laborious arithmetical calculations—a boon for engineers. One can see how Professor Charles Coulson, one of Maxwell's successors at King's, was doing no more than expressing a general view when he said of Maxwell: "There is scarcely a single topic that he touched upon which he did not change beyond recognition."[14]

CHAPTER THIRTEEN

GREAT GUNS

1863–1865

After a time of inner reflection, as was his way, Maxwell once more turned his conscious thoughts to electromagnetism. The result was a definitive work that will always stand as one of the world's greatest scientific achievements. He called the paper "A Dynamical Theory of the Electromagnetic Field," and, for once, Maxwell, the most modest of men, went so far as to toot his own horn. At the end of a long letter to his cousin Charles Hope Cay, he wrote:

> I also have a paper afloat, containing an electromagnetic theory of light, which, till I am convinced to the contrary, I hold to be great guns.[1]

In the first part of the paper he introduced to the world to a concept that had been Faraday's alone but was now also his—the field:

> The theory I propose may therefore be called a theory of the Electromagnetic Field because it has to do with the space in the neighbourhood of the electric or magnetic bodies and it may be called Dynamical Theory, because it assumes that in the space there is matter in motion by which the observed electromagnetic phenomena are produced.[2]

Even the most creative scientific theorists generally produce one great work on a subject and then move on. Maxwell was unique in the way he could return to a topic and raise it to new heights by taking a completely fresh approach. In his first paper on electricity and magnetism, he had used the analogy of an incompressible fluid to give mathematical expression to Faraday's concept of lines of force. In his second, he had built an entirely different imaginary model from spinning cells and idle wheels—a model that he admitted was "somewhat cumbersome"—but one that had yielded remarkable results. With it, he had not only accounted for all known electromagnetic effects but also had predicted two startling new ones: (1) displacement currents and (2) electromagnetic waves that traveled at the speed of light. Even the most enlightened of his contemporaries thought that the next step would be to refine this rather bizarre model, but, instead, Maxwell decided to put the model to one side and build the theory *ab initio* using only the principles of dynamics.

This was a fundamental shift in approach. He was no longer building imaginary models but rather trying to discern new scientific truth *directly* from the well-established mathematical relationships that were known as the laws of dynamics. These were the laws of motion that had been discovered by Newton, with one addition—the principle that energy was conserved in any closed system. The concept of energy in space was central to Maxwell's new approach in this paper, and he emphasized it by including the terms *dynamical* and *field* in the paper's title, distinguishing it from his previous paper "On Physical Lines of Force," where the focus was on forces.

He opened the paper by again outlining the differences between his own methods and those of theorists who assumed that forces could act at a distance without the aid of an intervening medium. He granted that action at a distance might seem, at first sight, to be the most natural mode of explaining the phenomena of electricity and magnetism, and he expressed his heartfelt respect for the work of men such as Weber but made it clear that he was looking beyond what seemed natural at first sight into something deeper. Weber's theory, he said, posed "mechanical difficulties" that "prevent me from considering this theory as an ultimate one."[3] This, it seemed, was Maxwell's modest way of implying that the new theory he was about to present might be, in a colloquial sense at least, the "ultimate one."

The mathematical laws of dynamics, everyone thought, belonged to material objects, and especially to machines with their levers, pulleys, gears, and springs. Maxwell's aim was to apply the same laws not to material bodies but to *the space that both contained and surrounded them* in electric or magnetic conditions—the field. He had already attempted this, with some success, in his spinning cells model, but only by constructing an imaginary machine that filled all space. The model had done its job by providing a kind of scaffolding that had given access to surprising new predictions—displacement currents and electromagnetic waves—but Maxwell now wanted to kick away the scaffolding and establish a theory that stood on its own, independent of any particular physical hypothesis.

As he saw it, *something* in space must be storing electromagnetic energy and transmitting its forces, and there was every reason to suppose that this medium, whatever its form, should obey the laws of dynamics, just as mechanical systems did. But how could he get a mathematical grip on this medium? The means came through his friends Thomson and Tait, who were in the early stages of collaborating on their great *Treatise on Natural Philosophy*, the first ever proper textbook on physics.[4] Part of their preparation was to study the writings of the great French mathematicians, whose work had been largely overlooked in Britain. Prominent among these was Italian-born Joseph Louis Lagrange, who had developed a formalized way of analyzing the motion of whole mechanical systems. Every system, no matter how large or complex, had a fixed number of independent modes of movement, and Lagrange had derived differential equations that showed how each of these was related to the whole system's kinetic and potential energy. The equations could be neatly lined up, like soldiers on parade, and solved to determine the system's motion for any set of starting conditions.[5]

A remarkable feature of Lagrange's method was that it treated the system as a "black box." Knowledge of the inputs and the system's general characteristics was enough to be able calculate the outputs; you didn't need to know the details of the internal mechanism. Characteristically, Maxwell found a compelling analogy to illustrate the point:

> In an ordinary belfry, each bell has one rope which comes down through a hole in the floor to the bellringer's room. But suppose that each rope, instead of acting on one bell, contributes to the motion of many pieces of machinery, and that the motion of each piece is determined not by the motion of one rope alone, but by that of several, and suppose, further, that all this machinery is silent and utterly unknown to the men at the ropes, who can only see as far as the holes above them.[6]

Nature's detailed mechanism could remain hidden, like the machinery in the belfry. As long as it obeyed the laws of dynamics, he should be able to derive the laws of the electromagnetic field without the need for any kind of model.

Nature's hidden mechanism was embodied in the *field*—the seat of energy in space—and, in Maxwell's Lagrangian formulation, the field became a coherent, connected system. It was, however, a system quite unlike anything seen or thought of before. The field was not a phantom: it held real energy that could be made to do mechanical work, and it exerted mechanical forces of electric and magnetic attraction and repulsion. Yet, for the most part, its components had an abstract quality. They were quantities that obeyed equations when given mathematical symbols but whose physical existence lay beyond anything we can detect with our senses.

Maxwell distinguished between two kinds of energy held by the field: electric energy was potential energy, like that in a coiled spring; and magnetic energy was kinetic, or "actual" energy, like that in a flywheel. To accommodate this energy, he assumed that all space, whether empty or occupied by material bodies, was packed with a medium that was capable of being set in motion and of transmitting that motion from one part of the field to every other part. To hold potential energy, the medium had a kind of electrical elasticity; and to hold kinetic energy, it possessed inertia and so acquired what he called "electromagnetic momentum" whenever it moved.

When acted on by an electric force, the elastic medium underwent distortion—Maxwell called it *displacement*—thus storing potential energy and exerting a spring-back force. The lines along which the distortion, or displacement, occurred were electric lines of force. (When Maxwell wrote in this paper of the number of lines of electric or magnetic force, he meant the number of *unit* lines of force, each being one unit of electric or magnetic flux.) The greater the distortion in any part of space, the greater was the density of the electric lines of force there, and the restoring force manifested itself in the world of ordinary matter as the tangible force of attraction or repulsion between electrically charged bodies. Any change in displacement constituted a brief electric current—the displacement current that he had described in his earlier paper. And the medium's momentum represented magnetic lines of force; the greater the momentum in any part of space, the greater the density of the lines there. Two more properties were needed. Where the medium occupied the same space as an ordinary substance—everywhere, that is, except in a vacuum—its elasticity and inertia were modified according to the ability of that substance to conduct, respectively, electric and magnetic lines of force.

The two systems of lines of force, represented by the medium's elasticity and momentum, were intimately linked by Maxwell's displacement current. Any change in the electric force was accompanied by elastic distortion in the medium. In the act of distortion there would be *some* movement in the medium, which implied momentum, and that represented magnetic force. So any *change* in electric force generated magnetic force. Moreover, the same thing happened in reverse: any *change* in magnetic force generated electric force. This two-way interaction was the final link in the connection between electricity and magnetism, and it was what gave rise to electromagnetic waves.

Maxwell also took account of two fundamental results from experiment: Ampère's finding that any loop of electric current acts as a magnet, and Faraday's finding that an electromotive force is generated in a circuit whenever there is a change in the number of magnetic lines of force that pass through it.

The medium connected with the world of ordinary matter at what Maxwell called "driving points," where real mechanical forces were exerted and real mechanical work was done, as in an electric motor or a generator. Any conducting circuit, for example, could be a driving point, or a driven point, or both at the same time. Every circuit was, as it were, geared to the medium by the lines of magnetic force that passed through it. These were the field's electromagnetic momentum, and the number of them that that passed through a circuit, or linked with it, determined how that circuit was geared to the rest of the field. The gear ratio, as it were, depended on the circuit's size, shape, and position.

Using his medium together with Lagrange's formulation of the laws of dynamics, Maxwell was able to calculate how every part of the field interacted with every other part.

We can get a rough idea of the process by taking a variant of his bell-ringers analogy. Imagine a long row of people pedaling exercise bicycles. None of the bicycles has the usual flywheel, but they all drive chains that run through holes in a wall and connect to the same unseen machinery (and hence, through the machinery, to all the other chains). Each rider has a different feeling through his pedals: to some they feel heavy; to others they feel light. Each is experiencing a different portion of the machinery's inertia through a feeling of weight in the pedals, and each is feeling some effect, though perhaps a tiny one, from every other rider's pedaling, his pedals being driven partly by his own efforts and partly by everyone else's. If one rider were suddenly to pedal harder, the effect would be felt to some degree by every other rider, though only after a delay while the change is transmitted through the medium—a delay so short as to be imperceptible to the riders.

Transferring the imagery back to Maxwell's theoretical reasoning, each rider's set of pedals becomes an electric circuit somewhere in space; the pedals' rate of rotation represents both the amount of current flowing in the circuit and the number of magnetic lines of force passing through it; the invisible machinery is Maxwell's all-pervasive momentum-carrying medium; and the chain linkage is the magnetic coupling of the circuit to that medium. Our exercise-bicycle analogy is partial—it doesn't illustrate how electrical effects are transmitted by means of the medium's elasticity—but Maxwell brought everything together and showed how the electrical and magnetic effects combined. Amazingly, the medium's properties of linked inertia and elasticity were enough to enable him to write equations that determined both the state of the field at any instant at a point in space and the physical forces exerted on any conducting circuits or charged bodies.

Maxwell's medium had electrical elasticity, and it also had the electromagnetic momentum that corresponded to magnetic lines of force. These properties were enough for him to calculate the speed at which disturbances were propagated through the medium, and he showed that this speed could be equivalently expressed as the ratio of the electromagnetic and electrostatic units of electric charge—the ratio that he had already identified, using his spinning cells model, with the speed of light. He had now shown, without the aid of any model, that the speed of light depends only on the elementary properties of electricity and magnetism and, moreover, that any electromagnetic wave, light included, consists of both an electric wave and a magnetic wave, always in phase, each vibrating at right angles to the direction of travel and at right angles to each other, as is illustrated in figure 12.2 in the previous chapter.

By using the Lagrangian formulation to arrive at the relevant equations, Maxwell had not only dispensed with the need for a mechanical model, but he had gone further. In his "Dynamical Theory" paper were the seeds of a truly revolutionary idea: some of nature's workings in the physical world not only do not need a mechanical model, but they *cannot* be explained in a mechanical way. For example, a current-carrying circuit "held" energy. This energy was real; it could be used in an electric motor to do mechanical work, but where *was* the energy? Not in the wire, but in the *field*—distributed through the surrounding space. It was *kinetic* energy, yet there was no evidence of movement. The magnetic lines of force that passed through the circuit constituted electromagnetic momentum, and this played a similar role to the familiar mechanical momentum, which was a body's mass multiplied by its velocity. But in electromagnetism, the momentum was disembodied; it was distributed throughout space. One can see why nineteenth-century scientists found it so difficult to take in these radical ideas: they had all been trained to think more in terms of things like colliding billiard balls that could be touched and measured.

Maxwell was doing nothing less than changing our concept of

reality. He was the first to recognize that the foundations of the physical world are imperceptible to our senses. All we know about them—possibly all we *can ever* know—are their mathematical relationships to things we can feel and touch. We may never understand what they *are*; we have to be content to describe them in an abstract way, giving them symbols and writing them in equations. As Freeman Dyson has aptly observed, Maxwell was in this way setting a prototype for the great triumphs of twentieth-century physics. Just as no one can truly picture Maxwell's electromagnetic momentum, so no one can visualize an electron, even though it can be rigorously defined in mathematical terms.

Maxwell had achieved the seemingly impossible—he had derived the theory of the electromagnetic field directly from the laws of dynamics. But, some may say, he did it only by postulating the existence of an all-pervading medium, or aether, a concept that Faraday refuted and that has since been discredited. The very idea of an aether seems preposterous to us today—how can a substance be so rarefied as to be imperceptible to the senses yet enough of an elastic solid, or in some versions an elastic fluid, to be able to transmit lateral vibrations at the speed of light? Though valid to a degree, such criticism largely misses the point.

Maxwell's theory was founded on facts, on the laws of electricity and magnetism that had been established by Faraday and others in experiments, and on the laws of dynamics that had been similarly proved. From these he made predictions—displacement currents and electromagnetic waves—that were later found by experiment also to be facts. Whereas other nineteenth-century physicists like William Thomson, Oliver Lodge, and George Francis Fitzgerald firmly believed in a material aether and gave defined mechanisms to their own versions of it, Maxwell gave his medium only properties. Though Maxwell had no idea of it, these properties turned out to be a prelude to the revelation by Einstein, in his special theory of relativity, of the fundamental properties of space and time. Maxwell had left his spinning cells model well behind and taken the endeavor to a new level. The field, with its intricately linked quantities that varied in space and time and were represented by abstract symbols, was to become the foundation for the great discoveries of the twentieth century, including the current theory of particle physics that is known as the Standard Model.

Maxwell wrote up his findings and published the paper in seven parts, including a twelve-page section on the electromagnetic theory of light.[7] When Maxwell introduced it at a presentation to the Royal Society in October 1864, the audience was bewildered; they simply didn't know what to make of it. A theory based on a bizarre model was bad enough, but one based on no model at all was beyond comprehension. One can sympathize with both Maxwell and his audience. It was a long and complex paper, difficult to summarize in a single talk and nearly impossible to assimilate quickly. Moreover, the mathematics was difficult. It described how the various quantities interacted with one another and how they varied in space and time. Most of the quantities, such as the electric and magnetic field intensities and flux densities, were represented by *vectors* that had both magnitude and a direction in three-dimensional space. Few people understood the mathematics of vectors, and what made things especially difficult for newcomers was that each vector equation came as a triple, one equation for each of the three dimensions. Maxwell's theory contained eight equations, but six of them were vector triples, so the total looked like twenty. One can see how it must have appeared impenetrable. The theory is usually presented today in the tight form of the four famous "Maxwell's equations," but Maxwell himself never summarized the theory in quite that way. He preferred to keep a more expansive arrangement, remarking that his eight equations might readily be condensed but that "to eliminate a quantity which expresses a useful idea would be a loss rather than a gain at this stage of our enquiry."[8] As usual, he was right: he had established a bridgehead in completely strange territory and it was prudent to keep all options for further advancement open. We'll see in later chapters how Oliver Heaviside arrived at the four equations now used by almost everyone.

There was, however, a further impediment, far more profound than a failure to follow the mathematics. William Thomson, a fine mathematician who had no difficulty absorbing the mathematics, expressed the view of many when he said that Maxwell had "lapsed into mysticism."[9] He and other fellows of the Royal Society were rooted in their Newtonian world, where every natural phenomenon had a mechanical explanation, and they failed utterly to see that Maxwell had opened the way to a new and different world. This was a historic moment. In his "Dynamical Theory" paper, Maxwell heralded one of the very rare events in science that the historian of science Thomas Kuhn has called a paradigm shift—a fundamental change in the set of shared beliefs and methods that guide scientists' thought and work—but, like most such changes, this one didn't properly take hold until several decades later, when a new generation of scientists with young and open minds had succeeded the old guard. As we will see, this process had its conflicts and unexpected twists.

The theory's construction had been an immense creative effort, sustained over a decade and inspired, from first to last, by the work of Michael Faraday. Thanks to Faraday's meticulous recording of his findings and thoughts in his *Experimental Researches in Electricity*, Maxwell had been able to see the world as Faraday did, and, by bringing together Faraday's vision with the power of Newtonian mathematics, to give us a new concept of physical reality, using the power of mathematics. But mathematics would not have been enough without Maxwell's own near-miraculous intuition; witness the displacement current, which gave the theory its wonderful completeness. The theory belongs to both Maxwell and Faraday.

Maxwell had put the "great guns" on parade, but it would be some time before they sounded. No one, probably not even Maxwell himself, recognized the full significance of his achievement, but he was by now a prominent figure in the scientific world, recognized and respected for his work on color vision and the kinetic theory of gases, and for his sterling efforts on electrical standards for the British Association for the Advancement of Science. His reputation brought much-appreciated kudos to King's College, and they lightened his burden by appointing a lecturer to assist with college duties. As to his own performance in the classroom, there is little reliable evidence, but the likelihood is that some of the problems that had plagued him at Aberdeen remained. Perhaps he had improved a little, and perhaps, as in Aberdeen, there were a few students like David Gill, who "could catch a few of the sparks that flashed as he thought aloud" and found him "supreme as an inspiration."[10]

Even with the extra help, Maxwell was finding it difficult to fit everything in. New ideas on kinetic theory and the theory of heat pressed on his mind, and he needed time to work them out. The work on electromagnetism, too, was far from finished. He wanted to write a substantial book, one that would bring much-needed order to the subject, help newcomers, and establish a solid base for his own further thinking. Another wish was to give more time to the estate and other local affairs at home, and to put in hand the grand enlargement to the Glenlair house that his father had designed and planned. Maxwell had thrived on the variety of work during the five years in London—college duties, home experiments, research on electromagnetism, and work for the British Association. He had also enjoyed the easy contact with fellow scientists; it was a joy simply being able to walk to meetings at the Royal Society and the Royal Institution. But he was still a country boy at heart and decided to resign his chair so that he and Katherine could take up a settled life in Galloway. The decision was made easier by knowing that he could safely hand the professorship to W. Grylls Adams, his very able assistant. Adams was younger brother to John Couch Adams, who, by discovering Neptune, had inspired the eponymous Adams Prize that Maxwell had won in 1857. The younger brother went on to have an illustrious career himself, becoming Astronomer Royal at Sydney. Maxwell agreed to help by returning to London in the winter and giving his usual course of evening lectures to working men.

In the spring of 1965, James and Katherine carefully packed the big color box once more for travel and left their Kensington house for Glenlair.

CHAPTER FOURTEEN

COUNTRY LIFE

1865–1871

When Maxwell asked Katherine to marry him, he had couched the request in a poem. It ran:

> Will you come along with me,
>   In the fresh spring tide,
> My comforter to be
>   Through the world so wide
> Will you come and learn the ways
>   A student spends his days
> On the bonny bonny braes
>   Of our ain burnside.[1]

This may not have been his finest piece of verse, but it wasn't meant for public scrutiny. It was an invitation to Katherine to share the home that meant so much to him: the ground that his parents had worked so hard to turn from stony scrub to pleasant and productive farmland; the hills and woods he had roamed as a boy with a walking staff his father had given him, alert to every movement of every creature that lived there. The "bonny bonny braes of our ain burnside" were those of the River Urr, which rippled brightly through the estate of Glenlair.

Carrying on in his father's tradition, Maxwell had steadily improved the estate. From the start, John Clerk Maxwell had intended to extend the modest house by building on a taller and grander section, but funds didn't match the ambition. Maxwell had spent many hours with his father discussing possibilities and making drawings. Now there was a chance of bringing the scheme to fruition. He went over the plans, changing and trimming where necessary to keep within what could be afforded, and he arranged for builders to start work in the following spring. Life was going well, but one day out riding Maxwell scraped his head on a tree. The cut seemed trivial but became infected, and he became dangerously ill. Once again, Katherine's devoted nursing saved his life. He was laid up for a month but then began to regain strength, and it wasn't long before they were riding again.

Maxwell remained shy with strangers, yet he left a strong impression on the memory of all he met; witness this recollection from somebody who met him in 1866:

> A man of middle height, with frame strongly knit, and a certain spring and elasticity in his gait; dressed for comfortable ease rather than elegance; a face expressive at once of sagacity and good humour, but overlaid with a deep shade of thoughtfulness; features boldly but pleasingly marked; eyes dark and glowing; hair and beard perfectly black, and forming a strong contrast to the pallor of his complexion. . . . He might have been taken, by a careless observer, for a country gentleman, or rather, to be more accurate, for a north country laird. A keener eye would have seen, however, that the man must be a student of some sort, and one of more than ordinary intelligence.[2]

And first impressions were amply borne out, as the same observer reports on further acquaintance:

> He had a strong sense of humour, and a keen relish for witty or jocose repartee, but rarely betrayed enjoyment by outright laughter. The outward sign and conspicuous manifestation of his enjoyment was a peculiar and brightness of the eyes. There was, indeed, nothing explosive in his mental composition, and as his mirth was never boisterous, so neither was he fretful or irascible. Of a serene and placid temper, genial and temperate in his enjoyments, and infinitely patient when others would have been vexed or annoyed, he at all times opposed a solid calm to the vicissitudes of nature.

One suspects that Maxwell would have been acutely embarrassed to hear any such analysis of his personal qualities. Serenity may have hidden internal struggles—he said he was as capable of wickedness as any man—but he rarely let inner thoughts escape. Introspection, he firmly believed, should never be performed in public. In an essay for the Apostles called *Is Autobiography Possible?* he had written:

> When a man once begins to make a theory of himself, he generally succeeds in making himself into a theory.
>
> . . . The stomach pump of the confessional ought only to be used in cases of manifest poisoning. More gentle remedies are better for the constitution in ordinary cases.[3]

His religion, too, was an intensely personal matter. Though now a trustee of his local parish and an elder in the Church of Scotland, he was not bound by any particular doctrine. Over the years, he refused several invitations to join the Victoria Institute, a body that aimed to establish common ground between science and religion. On the final occasion, in 1875, he gave his reasons:

> I think that the results which each man arrives at in his attempts to harmonize his science with his Christianity ought not to be regarded as having any significance except to the man himself, and to him only for a time, and should not receive the stamp of a society. For it is the nature of science, especially those branches of science which are spreading into unknown regions, to be continually [changing].[4]

Like Faraday and like Newton, Maxwell believed that God made the universe, that the laws of physics were God's laws, and that every discovery was a further revelation of God's great design. At the same time, as a devout Christian, he believed that the true nature of God was to be found in the Holy Bible, which he knew as well as any scholar of divinity. In his scientific work he treated all theories, including his own, as provisional until they had been backed by experimental results. How was this approach to be reconciled with a Christian faith that required absolute trust and belief in the absence of any material evidence? Much of the answer lay in how the scriptures were interpreted. It was no longer necessary to accept as literal truth the account in Genesis of God creating the world in seven days; that, and many other passages in the Bible, could be taken as metaphors. This process of interpretation wasn't easy, but for Maxwell it was necessary. He couldn't keep his science and religion in separate compartments—every crack between them had to be examined and repaired—but he found no formula to help with the task and thought himself no better qualified than anyone else to explain the connection between the spiritual and physical worlds.

Katherine was by now in her early forties, and it was clear that she and James would have no children. We don't know the reason but can be sure that it wasn't from choice: Maxwell greatly enjoyed playing with the children on the estate and, remembering the delights of his own childhood, loved to entertain them with tricks and games. There was also the matter of the succession. Without an heir, Glenlair would pass to a cousin on the Clerk side of the family, to whom it would be no more than a pleasant country estate. But one of Maxwell's mottos was that there is no use moping over what might have been. Outwardly, at least, he put the disappointment to one side, and made the most of life at is was.

The six years Maxwell spent at Glenlair were in no sense a time of retirement. He went to British Association meetings all over the country, sometimes acting as president of the Mathematics and Physics section, and he made annual visits to Cambridge, where, as examiner for the Mathematical Tripos, he did much to make the questions more interesting and more relevant to everyday experience. Meanwhile, his own work went on apace. The biggest project was his *Treatise on Electricity and Magnetism*. Here he set out not merely to present his own theory of electromagnetism but also to bring together everything that was known and to make this subject, still widely regarded as arcane, accessible to all scientists. It was a monumental feat—the book ran to almost a thousand pages and was eventually published in 1873.

The *Treatise* was a constant background task, but there was much else to do. His work for the British Association's committee on electrical standards had not stopped with the report on units and the production of the world's first standard of electrical resistance. Another difficult experiment was urgently needed—the measurement of the ratio of the electromagnetic and electrostatic units of electric charge. Much depended on the result because the speed of light, according to Maxwell's theory of electromagnetism, was exactly equal to this ratio. As we've seen, the ratio had already been measured experimentally by Kohlrausch and Weber, and their result, by Maxwell's interpretation, gave a theoretical value for the speed of light very close to the actual speed that had been measured by direct experiment. With so much at stake, it was important to confirm Kohlrausch and Weber's result by carrying out another experiment to measure the ratio of the units, preferably using a different method. Maxwell took on the job, this time in partnership with Charles Hockin, of St. John's College, Cambridge, and they carried out the experiment in London in the spring of 1868. By balancing the attractive electrostatic force between two oppositely charged metal discs against the repulsive electromagnetic force between two current-carrying wire coils, they measured the value of the ratio of the units of charge (which was also the speed of Maxwell's waves) at 288,000 meters per second, 7 percent below Kohlrausch and Weber's value and 8 percent below Fizeau's direct measurement of the speed of light. At first this seemed a disappointing result, but, on reflection, the experiment was a success. Maxwell's theory of electromagnetism had been strengthened because two independent experimental results now gave predicted wave speeds that, with reasonable allowance for experimental error, corresponded to the measured speed of light. We now know that the both the Fizeau and the Kohlrausch and Weber results were too high and Maxwell's too low, with the true value lying roughly midway between them.

As well as drafting most of his great *Treatise* while at Glenlair, Maxwell wrote another book, *The Theory of Heat*, and seventeen papers on a great variety of topics, each of which broke new ground. Most of this work lies outside our story, but one example will serve to demonstrate, from a different aspect, the almost-magical power of imagination that had enabled Maxwell to predict displacement currents and electromagnetic waves in empty space. In *The Theory of*

Heat Maxwell introduced what became, perhaps, his best known creation: Maxwell's Demon. The demon is an imaginary tiny creature who guards a hole in a partition between two gas-filled chambers. To start with, the gas in both chambers is at the same temperature. Temperature depends on the average speed of the gas molecules (strictly, on the average of the square of their speeds) but, according to Maxwell's own theory, at any given temperature some molecules will be going faster than average and some slower. At his hole, the demon operates a shutter, allowing only fast molecules to pass from the left chamber to the right one, and only slow ones to pass the other way. The gas in the right chamber gets hotter, and that in the left chamber colder—the demon is making heat pass from a cold substance to a hot one, thus defying the second law of thermodynamics. Maxwell was making the point that the law was not a physical one: it was statistical. To say that heat cannot flow from cold to hot was, as he put it, like saying that when you throw a glassful of water into the sea you can't get the same glassful out again. At the same time, he was posing a deep puzzle, and the demon lived up to its name, perplexing and intriguing generations of physicists. Among other things, it sparked the creation of information theory, which underpins our digital communications. Oddly, it was Maxwell's less frolicsome colleague William Thomson who named the demon; Maxwell wanted to call him a valve!

Whatever his business of the moment, Maxwell was apt to "feel the electrical state coming on"—thoughts on electricity and magnetism were never far away. His thoughts, like Faraday's, were often visual and no doubt included mental images of electric and magnetic forces and fluxes looping through space and embracing one another. These forces and fluxes were represented by vectors, mathematical entities that had both magnitude and direction. They also had a kind of three-dimensional geometry, but it was a very different geometry from anything in the textbooks of the time. It could be represented by equations, but these took a form that seemed arcane to most physicists and Maxwell looked for a way to demystify the subject. Was it possible to describe the geometry of vectors in a way that helped people to visualize the relationships between the physical quantities? Indeed it was. From his mental pictures, Maxwell coined three expressions that eventually became universal currency—*curl*, *divergence*, and *gradient*, the last two usually being abbreviated to *div* and *grad*. Maxwell originally proposed "convergence" and "slope," and, in his *Treatise*, replaced *curl* with the more formal "rotation," but, in essence, all the terms have stood the test of time. Once grasped, these images brought everything to life, and one senses that Faraday, with his acute visual imagination, would have recognized concepts that had already, if hazily, formed in his own mind's eye. They were concepts essential to the electromagnetic field.

Curl is the essence of the relationship between electricity and magnetism; it explains how the force of each connects with the flux of the other. At any given point in space, any vector, like magnetic flux or the velocity of wind in air, has a curl, which is itself a vector, though it may take the value of zero. Curl isn't easy to visualize, but it can be done. Think of water flowing in a river. The vector here is the speed and direction of flow, and, in general, it varies from point to point in the river. Now imagine a tiny paddle wheel somehow fixed at a point in the river but with its axis free to take up any angle. If (and only if) the water is flowing faster on one side of the paddle wheel than the other, the wheel will spin, and its axis will take up the position that makes it spin fastest. The curl of the water flow at out point is a vector whose magnitude is proportional to the rate of spin and whose direction is along the axis of spin, by convention in the direction a right-handed screw would move if it turned the same way as the paddle wheel. If the wheel doesn't spin, the curl of the water flow is zero. Curl is at the heart of two of the four equations in which Oliver Heaviside later summed up Maxwell's theory. At a point in empty space they say that the curl of the electric field force at a point is proportional to the rate at which the magnetic field force there is changing, and vice versa.

The river analogy also gives us an idea of divergence. Unlike curl, divergence is not a vector but a scalar—a term that mathematicians use to describe a quantity that has a magnitude (which can be positive, negative, or zero) but no direction. The divergence of the water flow at our fixed point is a measure of the excess of water flowing out of a small region surrounding the point compared with that flowing in. Assuming water is incompressible (very nearly true), the two amounts will be equal and the divergence zero. Unless, of course, we inject water in at our point, in which case the divergence will then have a positive value, or if we suck it out, in which case the divergence will be negative. The other two equations in which Heaviside summarized Maxwell's theory both employ divergence. At a point in empty space they say that the divergence of the electric field force and the divergence of the magnetic field force are both zero.

Gradient is a vector property of a scalar quantity. Imagine something that varies from place to place, like the height of land above sea level. Height is a scalar quantity, and its gradient at any given point is the slope of the land there along the direction of greatest incline (and is conventionally taken to point in the downhill direction). The gradient of an electric or magnetic potential is defined in similar fashion and manifests itself as the electric or magnetic field intensity, or force.

Along with curl, divergence, and gradient came another way of making the underlying physics clearer. Maxwell heard from his friend P. G. Tait about some strange mathematical entities called quaternions that represented rotations in three-dimensional space. They were the brainchild of the Irish mathematical genius Sir William Rowan Hamilton, namesake but no relation of the Sir William Hamilton who had taught Maxwell philosophy at Edinburgh University. Though he actually has much stronger claims to fame, Hamilton believed that quaternions were his greatest creation, and that they held the key to understanding all rotational phenomena in the physical universe. He had died in 1865, but not before recruiting a staunch disciple. Tait was bowled over by quaternions and became their vigorous champion. Not many followed his lead—quaternions were fearsomely complicated, and most people wanted nothing to do with them. For Maxwell, though, they presented an opportunity. Up to now he had written his various vector relationships as triple equations—one for each of the three space dimensions—but now he found that by using a form of quaternion representation, he could express the same relationships in single equations. Moreover, Hamilton had already built the mathematics of curl, divergence, and gradient into the quaternion system, so everything fit together beautifully. But very few people understood quaternions, and some of those who did understand them hated them, so Maxwell decided to play safe by including in his *Treatise* both the standard and the quaternion forms of the equations. A consequence was that he ran out of letters, having come close to exhausting the Roman and Greek alphabets! He then took what seemed to be the only course open and resorted to heavy Gothic Roman letters for his quaternion equations, so giving them a strange Teutonic air.

Thanks to quaternions, Maxwell could now present his theory of electromagnetism in eight equations rather than twenty, but it remained impenetrable to most physicists of the time. One can see why. Maxwell regarded the theory as work in progress and wanted to keep all the options for further advance open, even if this confused his contemporaries. In the *Treatise*, he still presented nearly all of his working in triple-equation format, with separate equations for each of the x, y, and z directions, and it was hard to see the forest for the trees. (To appreciate this difficulty, one need only look at figure 14.1, a diagram in which Maxwell used the notion of right-handed screws to help explain the x, y, and z components of the mechanical force on a current-carrying conductor in a magnetic field.)

Fig. 14.1. Maxwell's diagram using the notion of right-handed screws to help explain the components of force on a conductor in a magnetic field. (Used with permission from Lee Bartrop.)

Only the final equations appeared in the alternative quaternion format, as a kind of optional extra—one that most people preferred to do without. So things were to stay until six years after Maxwell's death, when Oliver Heaviside reduced the number of equations to four and replaced the quaternion representation with a much simpler kind of vector algebra. He thereby incurred the fury of Tait, who accused him of mutilating Hamilton's beautiful quaternions, but, as we'll see in a later chapter, Heaviside gave as good as he got—both of them were masters of literary invective and enjoyed a good scrap.

Early in 1869, Maxwell heard that James Forbes, the beloved mentor who had fired his interest in color vision and much else, had died. Maxwell felt the loss deeply, but Forbes's death meant that the post of principal at St. Andrews University was vacant, and a number of Maxwell's friends and associates urged him to stand for the job, among them Lewis Campbell, who was now professor of Greek at the university. Maxwell was at first reluctant, commenting that "my proper line is in working, not governing, still less in reigning and letting others govern."[5] However, he did believe passionately in good education, and his supporters were so enthusiastic that he began to feel that maybe he could do some good in the position. Eventually, he was persuaded to let his name go forward.

It was a politically charged appointment, but Maxwell knew nothing of politics and wrote, rather pathetically, to a London acquaintance:

> I have paid so little attention to the political sympathies of scientific men that I do not know which of the scientific men I am acquainted with have the ear of the Government. If you can inform me, it would be of service to me.[6]

Given such candid naivety, it is not surprising that Maxwell didn't get the job. Perhaps this was just as well—one thinks of a goldfish about to enter a tank of piranhas—but with Maxwell one can't be sure; for all we know, he might have transcended the political squabbles. The job went to the university's professor of humanity, John Campbell Shairp, an eminent scholar of poetry and a supporter of the newly forged Liberal Party, which had just come into power.

After he lost his job at Aberdeen, Maxwell had been rejected for the chair of natural philosophy at Edinburgh University, but, shortly afterward, he was accepted for a similar post at King's College, London. Now, ten years later, the sequence was repeated, and once again Scotland's loss turned out to be England's gain. Shortly after St. Andrews had turned him away, Cambridge University asked Maxwell to accept an important new professorship in experimental physics. The university's chancellor, the duke of Devonshire, had offered a large sum toward the setting up of a new department and the building of a grand new laboratory. The task of getting the laboratory designed and built would fall to the first professor. Ideally, the university authorities wanted a first-rank scientist who had experience in running a leading research and education establishment. William Thomson was the obvious choice, but he didn't want to leave Glasgow, where he had painstakingly built up a fine research center, starting in a converted wine cellar. Second on Cambridge's list was Hermann Helmholtz, but he had just accepted a top post in Berlin and wanted to stay there. Maxwell was the third choice, but a popular one among the younger fellows. Their spokesman was J. W. Strutt, who later, as Lord Rayleigh, succeeded to the same post, and he put their case fervently to Maxwell, imploring him to come.

Once again, Maxwell was initially reluctant. He loved his home and had settled at Glenlair into a good life that combined laird's duties with scientific work in a pleasant and fruitful way. But he had affection for Cambridge, too, and saw what a rare opportunity the new project offered for the university and, indeed, the country. Was he really the man for the job, though? He wasn't sure. At length he accepted, on condition that he could leave after a year if he wished. This didn't imply any lack of commitment. As in all things, he was determined to do his best, but, being acutely aware of his lack of experience in directing a big and complex operation, he wanted to be free to withdraw if he found that he wasn't able to run the show well. In March 1871, twenty-one years after entering Cambridge University as a student, he was appointed its first professor of experimental physics. Once again, he and Katherine moved south.

CHAPTER FIFTEEN

THE CAVENDISH

1871–1879

The first task at Cambridge was to draw up detailed specifications for the new laboratory building and its equipment. Everything had to be at the forefront of scientific progress, yet it was important to avoid mistakes that would be expensive or impossible to put right later, so Maxwell visited the best university laboratories in the country—the newly built Clarendon Laboratory at Oxford and William Thomson's laboratory at Glasgow—to learn all that he could from their experiences. There needed to be abundant light, plenty of space for bulky apparatus, a battery room, and a tower tall enough to provide water pressure to drive a powerful vacuum pump. To enable delicate experiments to be done on magnetism, one room needed to have table surfaces that were insulated from ordinary vibrations in the building, so Maxwell specified stone-topped tables that were four feet square, to be supported directly from the building's foundations by brick piers that came up through the floor without touching it. The job of producing a design to meet these and a host of other requirements was assigned to the young local architect William Fawcett, who had studied at Jesus College.

Sessions with Fawcett no doubt took Maxwell back to the hours spent with his father mulling over plans for the extension to Glenlair. The current project was on a far grander scale, but the design process—outline sketches, discussion, fuller sketches, and, eventually, detailed drawings—was probably much the same. The outcome of these sessions was a fine building, plain and functional yet in keeping with the much-older buildings of the rest of the university. It seemed to exude confidence about its place in the scheme of things and went on to serve Cambridge well for over one hundred years, becoming the site of many important discoveries, such as those of the electron and the structure of DNA.

Plans were agreed and work got under way. Meanwhile, there were lectures to give and Maxwell had nowhere to call his own. He wrote to Lewis Campbell:

> I have no place to erect my chair, but move about like the cuckoo, depositing my notions in the Chemical lecture room 1st term; in the Botanical in lent, and in Comparative Anatomy in Easter.*

There was also, of course, the customary inaugural lecture, Maxwell's third. Through some misunderstanding, a group of senior professors came to his first ordinary lecture to undergraduates, believing it to be the formal one, and Maxwell, with a twinkle in the eye, solemnly explained to them the difference between the Fahrenheit and Centigrade scales of temperature. In the real inaugural lecture, he developed themes that he had already expounded in Aberdeen and London: his job was to teach students to think for themselves, to seek out truth, and to recognize and expose falsity in all its forms. He also emphasized once more the essential role of practical work in science. One passage clearly evokes Faraday:

> When we shall be able to employ in scientific education, not only the trained attention of the student, and his familiarity with symbols, but the keenness of his eye, the quickness of his ear, the delicacy of his touch, and the adroitness of his fingers, we shall not only extend our influence over a class of men who are not fond of cold abstractions, but, by opening at once all the gateways of knowledge, we shall ensure the association of the doctrines of science with those elementary sensations which form the obscure background of all our conscious thoughts, and which lend a vividness and relief to ideas, which, when presented as mere abstract terms, are apt to fade entirely from the memory.

And, in a further passage that might have been composed by Faraday himself, Maxwell said:

> We may find illustrations of the highest doctrines of science in games and gymnastics, in travelling by land and by water, in storms of the air and of the sea, and wherever there is matter in motion.

Building progress was reasonable but seemed frustratingly slow to everyone waiting to get experiments under way. Even Maxwell's patience was tested by the gas men, whom he described as "the laziest and the most permanent of the gods that have been hatched under heaven." At last all was ready and the laboratory opened in the spring of 1874. It was to have been called the Devonshire, after its founder, but shortly before the inauguration a decision was made to call it instead the Cavendish Laboratory. This way, the name would commemorate not only the duke, whose family name was Cavendish, but also his great uncle Henry Cavendish, one of the greatest British scientists. Henry Cavendish was a very strange character. Extremely shy, he had lived as a recluse, venturing out only occasionally to scientific meetings and communicating with his domestic staff by written notes. Women servants were fired if they allowed themselves into his sight. He rarely spoke. An acquaintance said: "He probably uttered fewer words in the course of his life than any man who lived to four score years, not at all excepting the monks of La Trappe." His genius lay in performing amazingly accurate experiments using simple but brilliantly effective apparatus of his own design. With his faithful servant Richard as laboratory assistant, he had achieved remarkable experimental results, for example, proving that water was not an element but a compound, and measuring the density of Earth to within 2 percent of its correct value. He had also played an important, though indirect, part in establishing the career of Michael Faraday—along with Count Rumford he had been one of the founding fathers of the Royal Institution.

Henry Cavendish had been shy even of publishing. Some of his work had been published, but much had not, and at around the time of the laboratory's inauguration the duke handed Maxwell a great pile of papers—his great uncle's accounts of electrical experiments from 1781 to 1791—with a request to edit them for belated publication. No doubt Maxwell's heart sank—he already had more than enough to do—but he undertook to look through the papers. He could hardly refuse the university's great benefactor, but it wasn't entirely a forced decision. He had a high regard for the duke, who, like himself, had been second wrangler and first Smith's Prize man, and they shared a deeper bond—both felt passionately about the importance of practical work in scientific education. The duke's generous gift of a new laboratory had not been hailed with unalloyed joy in the country, or even in Cambridge. Skeptics and cynics abounded. Even the progressive scientific journal *Nature* commented disparagingly that, with luck, the laboratory might in ten years reach the standard of a provincial German university. And many thought that although scientific research was a necessary activity, demonstration experiments were pointless. Among them was the celebrated Cambridge tutor Isaac Todhunter. One day, Maxwell bumped into Todhunter in the street outside the laboratory and asked him to come in to see an example of conical refraction, a phenomenon that was much talked about but rarely witnessed because it was so hard to set up. Todhunter replied: "No thank you. I've been teaching it all my life and I don't want my ideas upset by seeing it now."

Maxwell was no politician but saw clearly how important it was for the laboratory to establish a reputation with some early successes. From our distant perspective it may seem surprising that he didn't straightaway set about trying to verify his own theory of electromagnetism by detecting displacement currents or electromagnetic waves, but such experiments were too difficult and risky for the purpose. He set up instead a research program focused mainly on high-precision measurements of fundamental physical quantities. This was unspectacular but important work and it brought in solid results. One such experiment was to verify Ohm's law, which said that no matter what amount of current flowed in a conductor, the ratio of current to voltage remained constant. Maxwell's student, George Chrystal from Aberdeen, vanquished the doubters by demonstrating that the law held true for a vast range of currents within one part in a million million.

Political considerations aside, Maxwell saw his job as helping science to advance on a broad front. He had many interests besides electromagnetism and, in any case, had no thought of establishing a "Maxwell school." Defying the gloomy predictions, he had no difficulty in attracting a highly talented set of young researchers, some of whom had left good jobs elsewhere to come and work with him. His style was not at all dictatorial—everyone was encouraged to come up with his own ideas and solve his own problems—but Maxwell kept a fatherly eye on progress, and advice from one of the greatest scientific minds of all time was dispensed with unfailing generosity and humor. His students loved him, and many went on to have distinguished careers elsewhere. For example, Richard Glazebrook was the first director of the National Physical Laboratory of Great Britain; Donald MacAlister became president of the General Medical Council and principal of Glasgow University; William Napier Shaw became known as the father of modern meteorology; and Ambrose Fleming became Guglielmo Marconi's right-hand man and invented the thermionic valve. There was also John Henry Poynting, who, as we'll see, made an important contribution to the theory of electromagnetism. Funds were tight, but Maxwell donated his own equipment to the laboratory and bought several hundred pounds' worth of new apparatus from his own pocket during his tenure.

Maxwell's *Treatise on Electricity and Magnetism* was published in 1873. Still in print, it is probably, after Newton's *Principia Mathematica*, the most renowned book in physics. Before the *Treatise*, students and scholars had no substantial books to help with their studies, just scattered writings. Now they had a book that covered everything. In a thousand crisply written pages, Maxwell set out all that he knew. At first sight, it looks like a textbook—and, indeed, most modern textbooks are derived from it—but it was drawn up many years before its subject matter became standard in university courses, and it was really more of an explorer's report, written for those who wanted not only to follow but to venture further. This audience included himself: Maxwell was still exploring and was partway through an extensive revision of the book at the time of his death. The subject matter was complex, difficult, and new, so it is not surprising that early readers found it hard going. Like all of Maxwell's work, it was thorough: it covered not only theory but also practical applications, for example, setting out how to make galvanometers and the procedure to be followed when correcting compass readings on iron ships. It was emphatically *not* a showcase for Maxwell's own theory of electromagnetism. The theory was there, but you had to hunt for it. The topic of electromagnetism is not introduced until article 475 in volume 2, and the sublime creation on which his whole theory depends, the displacement current, is slipped in with no fanfare in article 610. This can be fairly interpreted as an example of Maxwell's excessive modesty, but it is also a reminder to us that the *Treatise*, even now a wonderfully authoritative text on electricity and magnetism, was written at a time when Maxwell's own theory was only a contender, generally less favored than Weber's action-at-a-distance theory. And although the theory was wholly encapsulated in its equations, it was then like a vehicle chassis without a body—for example, it predicted electromagnetic waves but gave no indication of how they might be produced or detected in a laboratory. We'll see in the next chapter how a few men from the next generation were able to see the latent power and beauty of the theory, to take it further, and to cast it in a cogent form that came to command universal assent.

Only after several decades did the *Treatise* begin to achieve the acclaimed status it holds today. Its early readers faced more obstacles than those inherent in Maxwell's own theory of the electromagnetic field. To many, the whole of electricity and magnetism was an arcane topic on the fringes of known science, and they had to wrestle with what seems today to be an idiosyncratic selection and arrangement of topics. For example, as early as article 8 on page 7, the reader is faced with the heading "Discontinuity of a Function of More Than One Variable" and presented with a difficult equation. And most of the first third of the book is taken up with a minutely detailed and highly mathematical account of electrostatics. What really brought the *Treatise* to prominence were the rapid advances in electrical communications and in electrical power and machines in the late nineteenth and early twentieth centuries. Technology had forged ahead: it needed trained scientists and engineers, and those charged with the training found a superb text in Maxwell's *Treatise*. They didn't need the long chapters on such things as spherical harmonics, but Maxwell had already set out very clearly much of what they *did* need. The *Treatise* was ahead of its time.

Henry Cavendish had also been ahead of his time. When Maxwell looked through Cavendish's accounts of electrical experiments performed a hundred years earlier, he was astonished. It was like finding a dozen unpublished plays by Shakespeare. Among a string of stupendous results, Cavendish had demonstrated the inverse-square law for the force between electrical charges more effectively than Coulomb, after whom the law was named. He had also discovered Ohm's law fifty years before Ohm and twenty years before Volta produced the first electric battery. His method was simple and painful. He connected two wires to the oppositely charged parts of a Leyden jar and grasped both wires in one hand. He then repeated the procedure with various circuit arrangements, each time judging the strength of the current by measuring how far up his arm he could feel the shock. One day, a distinguished American, Samuel Pierpoint Langley, visited the Cavendish and was horrified to find Maxwell and some of the students with sleeves rolled up, repeating the experiment. He declined an invitation to join in and remarked: "When an English man of science comes to the United States we do not treat him like that."

Maxwell could have delegated the huge task of editing Cavendish's papers, but he decided to take on the job himself, "walking the plank" with them, as he put it to William Thomson. In retrospect, it seems extraordinary that Maxwell chose to spend time on this work when he could have spent it on his own research. He, of course, saw it differently. His own ideas on electromagnetism and other topics were still developing—"decocting" in the subconscious—and he didn't know that he had only five years to live. Cavendish's work was an important part of scientific history, and it needed to be presented in a proper way. Maxwell went to great trouble to write an interesting and informative narrative, even checking such details as whether the Royal Society premises in the 1770s had a garden.

As always, Maxwell took everything in his stride. Along with editing the Cavendish papers and repeating many of the experiments, he took on the scientific editorship of the ninth edition of the *Encyclopaedia Britannica* jointly with T. H. Huxley. He also wrote some brilliant and original papers, refereed many more by other authors, reviewed books, and wrote another of his own. *Matter in Motion* is a pedagogical gem. In only 122 pages, it explains the principles of dynamics with crystal clarity in plain language with no more than a few equations. Yet nothing is dumbed down: the reader is required to think. In contrast, his paper "On Boltzmann's Theorem on the Average Distribution of a Number of Material Points" contained some of his most complex mathematics. It laid the foundations for the development by Ludwig Boltzmann and Josiah Willard Gibbs of statistical mechanics, a difficult but useful set of methods by which the properties of a substance can be deduced from the behavior, en masse, of its molecules.

Though still shy with strangers, Maxwell left an impression on everyone he met. Lewis Campbell described the effect:

> One great charm of Maxwell's society was his readiness to converse on almost any topic with those he was accustomed to meet . . . no one talked to him for five minutes without having some perfectly new ideas set before him; some so startling as to confound the listener, but always such as to repay a thoughtful examination.

The only direct record we have of Maxwell's spoken words is from the text of his formal lectures, and most of these, by their nature, give us only the faintest hint of what his companions enjoyed every day. Perhaps one extract can take us a little closer. When giving a public lecture about Alexander Graham Bell's new invention, the telephone, he talked of Bell's father, who had lived in Edinburgh and was an expert in elocution. In his still strong Galloway accent, Maxwell told his audience:

> His whole life had been employed in teaching people to speak. He brought the art to such perfection that, though a Scotchman, he taught himself in six months to speak English. I regret extremely that when I had the opportunity in Edinburgh I did not take lessons from him.[1]

Maxwell was rarely seen in Cambridge without a dog, and his terrier Toby was quite at home in the laboratory. Lewis Campbell reports that Toby always became uneasy when he heard electric sparks, but when Maxwell called him to his station, he would sit down between his master's feet and allow the sparks to be applied to his back, growling all the time in an odd manner but not showing any real signs of discomfort. There wasn't an ounce of cruelty here—Campbell reports elsewhere in his book of his friend's way with animals and love of all creatures.[2]

Maxwell enjoyed Cambridge life, just as he had done as a student. When time allowed, he attended an essay club that was like a senior version of the Apostles. In one essay he challenged the widely held belief that scientific laws implied a mechanical universe whose whole future is predictable, given sufficient knowledge of its present state. In doing so, he gave a remarkable outline of chaos theory—a hundred years before mathematicians began to develop the subject:

> When the state of things is such that an infinitely small variation of the present state will alter only by an infinitely small quantity the state at some future time, the condition of the system, whether at rest or in motion, is said to be stable; but when an infinitely small variation in the present state may bring about a finite difference in the state of the system in a finite time, the condition of the system is said to be unstable.
>
> It is manifest that the existence of unstable conditions renders impossible the prediction of future events, if our knowledge of the present state is only approximate and not accurate.[3]

Maxwell was a relative newcomer to the senior ranks at the university, but his influence soon spread well beyond his own department. People who were at first hostile or indifferent to the new laboratory began to recognize its achievements in research and were won over by Maxwell's enthusiasm and unaffected charm. In only a few years, the Cavendish had established itself, and science at Cambridge had entered a new age and set an example for other universities to follow in experimental physics.

The Maxwells lived in a comfortable house in Scroope Terrace, close to the laboratory. Katherine began to suffer from poor health and was for a while seriously ill, though the condition was never properly diagnosed. It was James's turn to be the nurse; he slept in a chair at her bedside for three weeks, yet carried on his work at the laboratory with apparently undiminished vigor. Perhaps because of her illness, Katherine wasn't always welcoming to Maxwell's colleagues and students, so he sometimes conducted business at the laboratory that might have been done more congenially over a cup of tea at home. One can see how Katherine acquired her reputation as a "difficult" woman. We will never have the full picture, but it is clear that, whatever the tensions, James and Katherine remained unswervingly devoted to one another.

Four months of every year were spent at Glenlair. Maxwell wrote many of his articles and reviews there and kept in touch with those working in the laboratory during the vacations. For some of the short summer courses, he allowed female students to attend—a remarkable departure for Cambridge. When the Maxwells were at Glenlair, the postman from the Kirkpatrick Durham post office was kept busy, as usual, and many of the packages contained proofs to be checked and corrected. Publishers became the second occupational group, after gas men, to arouse Maxwell's wrath. They seemed determined to economize by cutting every corner, and he reckoned their maxim must be "a stitch in nine save time."

In 1877, Maxwell had begun to suffer from heartburn. For a year and a half, the problem was no more than a nuisance—he ran the laboratory, gave his lectures, and wrote his articles, all with his usual sparkle. But then colleagues began to notice a loss of spring in his step and Maxwell did something he had never done before. He turned down a request to write an article for T. H. Huxley's *English Men of Science*, pleading overwork. In 1879, he and Katherine went to Glenlair as usual for the summer, hoping that he would recover with rest. By September, Maxwell was getting violent pains, but he insisted that a visit by his assistant William Garnett and his wife should go ahead as planned. Garnett was shocked at the change in Maxwell's appearance but marveled at the care he still gave to his guests and at the way he still conducted prayers each evening for the whole household. Maxwell showed Garnett the oval curves and other memorabilia from his childhood, and walked with him down to the river, pointing out where he used to swim and to sail in the washtub. This was the longest walk he had taken for weeks. When Katherine took the guests for a drive, he couldn't go with them because the shaking of the carriage was unbearably painful.

Maxwell suspected that he had contracted the same type of abdominal cancer that had killed his mother at the same age. They sent for a specialist, Dr. Sanders, from Edinburgh. He arrived on October 2 and confirmed the worst: Maxwell couldn't expect to live much longer than a month. Sanders suggested he travel to Cambridge, where Dr. Paget, an expert in palliative care, would be able to make his last few weeks as bearable as possible for both Katherine and James. Fortunately, Katherine was having a respite from her own illness and was able to organize packing and travel. On arrival at Cambridge, Maxwell was barely able to walk the few yards from the train to a carriage, but once in the care of Dr. Paget, his pain was much relieved and for a few days he felt better. Word spread among friends and colleagues, and there was even some hope that he might recover. Such hopes were soon dashed. His remaining strength began to ebb, and it was clear to all that he was dying. Dr. Paget later described this time:

> As he had been in health, so was he in sickness and in the face of death. The calmness of his mind was never once disturbed. His sufferings were acute for some days after his return to Cambridge, and, even after their mitigation, were still of a kind to try severely any ordinary patience and fortitude. But they were never spoken of by him in a complaining tone. In the midst of them his thoughts and consideration were rather for others than for himself. Neither did the approach of death disturb his habitual composure. . . . A few days before his death he asked me how much longer he could last. This inquiry was made with the most perfect calmness. He wished to live until the expected arrival from Edinburgh of his friend and relative Mr. Colin Mackenzie. His only anxiety seemed to be about his wife, whose health had for a few years been delicate and had recently become worse. . . .

> His intellect also remained clear and apparently unimpaired to the last. While his bodily strength was ebbing away to death, his mind never wandered or wavered, but remained clear to the very end. No man ever met death more consciously or more calmly.

Maxwell's local doctor at Glenlair, Dr. Lorraine, had written to Dr. Paget with notes on the patient's condition. This, of course, was standard practice, but there was more. Dr. Lorraine had such admiration for his patient that he included a spontaneous tribute.

> I must say he is one of the best men I have ever met, and a greater merit than his scientific achievements is his being, so far as human judgement can discern, a most perfect example of a Christian Gentleman.

He was, indeed, one of the best men that anyone could meet, a genius without vanity, someone who made people feel good about themselves and the world in general. His own reflections on his life were characteristically modest. He told his friend and Cambridge colleague Professor Hort:

> What is done by what I call myself is, I feel, done by something greater than myself in me. . . . I have been thinking how very gently I have always been dealt with. I have never had a violent shove in all my life. The only desire which I can have is like David to serve my own generation by the will of God, and then fall asleep.

James Clerk Maxwell died on the November 5, 1879. Katherine and his friend and cousin Colin Mackenzie were at his bedside. The next Sunday, a memorial service was held at St. Mary's Church in Cambridge. The task of giving a voice to the all-pervading sense of loss fell to the Reverend H. M. Butler, one of Maxwell's friends from his student days, who was now headmaster of Harrow School. He chose his metaphor well:

> It is not often, even in this great home of thought and knowledge, that so bright a light is extinguished as that which is now mourned by many illustrious mourners, here chiefly, but also far beyond this place.

Maxwell's old schoolfellow P. G. Tait echoed this thought, adding a typically combative touch of his own. He wrote in *Nature*,

> I cannot adequately express in words the extent of the loss which his early death has inflicted not merely on his personal friends, on the University of Cambridge, on the whole scientific world, but also, and most especially, on the cause of common sense, of true science, and of religion itself, in these days of much vain-babbling, pseudo-science, and materialism. But men of his stamp never live in vain; and in one sense at least they cannot die. The spirit of Clerk Maxwell still lives with us in his imperishable writings, and will speak to the next generation by the lips of those who have caught inspiration from his teachings and example.

Maxwell's body was taken to Glenlair and buried next to those of his father and mother in Parton churchyard. Katherine was buried there seven years later, and the four share a headstone. A simple plaque by the roadside in front of the church now describes his life and achievements, and it concludes:

> A good man, full of humour and wisdom, he lived in this area and is buried in the ruins of the old Kirk in this Churchyard.

A few miles away, the ruin of the house of Glenlair stands with empty windows and roofless gables, having been destroyed by fire in 1929.[4]

Butler and Tait had not exaggerated the sense of loss inflicted by Maxwell's early death. He had been in full flow, and who knows what else he might have gone on to do. But he has been an inspiration to physicists and engineers ever since. Perhaps more than any other scientist's, his personality comes over in his work: he seems to elicit a unique blend of wonder and affection. An editorial in the *Times Educational Supplement* in 1925 summed this aspect up very well. It said:

> To scientists, Maxwell is easily the most magical figure of the 19th century.

In the story of the electromagnetic field, Maxwell was a lone actor in his time, just as Faraday had been in his. Not until the following generation did anyone else truly understand what Faraday and Maxwell had been trying to tell them. The way was then led by a small band of individuals with disparate but complementary talents who came to be called the "Maxwellians." Oliver Heaviside was one of these. As we will see, he was a prickly individual whose criticism could be withering, but when he wrote of Maxwell he seemed to have joy in his heart:

> A part of us lives after us, diffused through all humanity, more or less, and all through nature. This is the immortality of the soul. There are large souls and small souls. . . . That of a Shakespeare or Newton is stupendously big. Such men live the best part of their lives after they are dead. Maxwell is one of these men. His soul will grow for long to come, and hundreds of years hence will shine as one of the bright stars of the past, whose light takes ages to reach us.[5]

CHAPTER SIXTEEN

THE MAXWELLIANS

1850–1890

"One scientific epoch ended and another began with James Clerk Maxwell."

—Albert Einstein

"From a long view of the history of mankind—seen from, say, ten thousand years from now—there can be little doubt that the most significant event of the nineteenth century will be judged as Maxwell's discovery of the laws of electrodynamics."

—Richard P. Feynman

Einstein's and Feynman's words aptly convey the momentous effect that James Clerk Maxwell's theory of the electromagnetic field has had on science and technology, and indeed on human history. But scientific theories rarely spring fully formed from the minds of their originators. It often happens that a following generation of scientists has to refine and codify a theory before it becomes assimilated into the common body of scientific knowledge—a process that can take decades. So it was with Maxwell's theory.

Although Maxwell had set out the theory as clearly as he could in his paper "A Dynamical Theory of the Electromagnetic Field" and later in his *Treatise on Electricity and Magnetism*, almost nobody understood it during his lifetime. Not only was the mathematics difficult, but its whole approach was based on Michael Faraday's theoretical vision, which still seemed bizarre to most physicists. Maxwell's seminal work on another topic—the statistical properties of matter—

was being taken forward by two men of genius, Ludwig Boltzmann and Josiah Willard Gibbs, but when Maxwell died, his theory of electromagnetism sat for a while like an exhibit in a glass case, admired by some but out of reach.

Maxwell himself made no attempt to verify the theory experimentally while at the Cavendish. This is often attributed, with good grounds, to his self-effacing modesty, but, as we have seen, there was a further reason: the new laboratory needed to establish a reputation with some early successes, and experiments to find displacement currents or electromagnetic waves would have been too risky for the purpose. By the end of the 1870s, the laboratory's reputation had been secured, and one might have expected Maxwell's successor, Lord Rayleigh, to take up the challenge—he was a great admirer of Maxwell, and ten years earlier, as John William Strutt, he had been one of the young fellows who had begged Maxwell to accept the post. Rayleigh, though, had his own priorities as director of the laboratory. The first was to set the enterprise on a sound financial footing. Maxwell had been reluctant to pump the duke for more money, but Rayleigh, a hardheaded man of business as well as a scientist, had no such compunction. At length, the duke agreed, Rayleigh put in some money of his own, and the laboratory acquired the equipment it needed to build on its early successes. Rayleigh also introduced systematic training in laboratory techniques, moving on from Maxwell's laissez-faire approach, and he needed what time was left for his own research—among other outstanding achievements, he discovered argon and explained how scattering of light makes the sky appear blue. Whether from want of interest or from want of ideas, neither Rayleigh nor any of his researchers at the Cavendish made a serious and sustained attempt to confirm or develop Maxwell's theory. When the advance came, it was not from Cambridge.

Oliver Heaviside was born in 1850, the youngest of four sons in a respectable but poor family living in London. Scarlet fever at the age of eight left him partially deaf, and he found himself excluded from street games because he couldn't hear what the other boys were saying. He was thrown on his own resources and began to build a defense against the barbs of the world. A stubborn independence took hold of him, almost against his will, and held its grip until the day he died. He did well at school, despite defying its rote-teaching methods, but university was beyond the family's resources, and instead he spent two years studying on his own at home, reading everything he could find on scientific topics. This privilege hadn't been extended to his brothers, who were working and contributing to the household income: Oliver's uncle by marriage, Charles Wheatstone, had interceded on his behalf. This is the same Wheatstone who, twenty years before, had fled from giving a lecture at the Royal Institution, and thereby caused Faraday to give his impromptu talk on "Ray-vibrations." Wheatstone had also, with his partner William Fothergill Cooke, built the first commercial telegraph in Britain and expanded the business at a prodigious rate. With Wheatstone's recommendation, Oliver got his first (and only) job, at the age of eighteen as a telegraph operator with the Danish-Norwegian-English Telegraph Company at the excellent salary of £150 a year.

The company had just laid its first North Sea cable, and Heaviside was posted to its main Danish operating station at Fredericia. He was soon enraptured by the telegraph and the mystery of how it worked. The equipment used visual cues rather than sound, so his partial deafness was no handicap. He quickly mastered Morse code, but the part of the job he really enjoyed was fixing faults. Suboceanic telegraphers were the technological elite of the day; operators were free to experiment with advanced equipment like rheostats, bridges, shunts, and condensers; indeed, they had to, simply to keep the traffic flowing. Heaviside became a star troubleshooter, though for him each problem was not merely something to be fixed but a means of probing the strange ways of electricity, which often baffled even the most experienced of his colleagues. At the age of twenty, he was posted to the company's English headquarters at Newcastle, given an increase in salary, and promoted to chief operator.

He continued to shine, on one occasion saving the company money by accurately locating a mid-ocean fault by some deft measurements and calculations before the cable-repair ship left shore. In his free time, he continued his studies into electricity and mathematics, and he began to write scientific papers. One day, he opened a book in Newcastle's public library and was captivated. From that moment, the course of his life was set. Many years later, he recalled the experience:

> I remember my first look at the great treatise of Maxwell's when I was a young man. Up to that time there was not a single comprehensive theory, just a few scraps; I was struggling to understand electricity in the midst of a great obscurity. When I saw on the table in the library the work that had just been published (1873) I browsed through it and was astonished! I read the preface and the last chapter, and several bits here and there; I saw that it was great, greater and greatest, with prodigious possibilities in its power. I was determined to master the book and set to work.[1]

Heaviside's vocation was to find out all he could about electricity and spread the word. It would be a full-time job. His work for the telegraph company was now becoming routine and tiresome, with long hours spent sending and receiving telegrams, so at the age of twenty-four he resigned his post and returned to his parent's house in London to work as a full-time unpaid researcher. He could expect only a pittance from writing articles and books, but that didn't bother him. He was doing what he saw as his duty: if society chose to reward him generously for his work, that would be fine; if not, his parents and brothers would be doing *their* duty by supporting him. It was a near-solitary life, but he had no feeling of loneliness. As he later explained:

> There was a time indeed in my life when I was something like old Teufelsdröckh in his garret, and was in some measure satisfied with a mere subsistence. But that was when I was making discoveries. It matters not what others think of their importance. They were meat and drink and company to me.[2]

One question in particular fascinated him: exactly how do electrical signals travel along wires? The transmission line became a lifetime study, and he made it his own—the theory of transmission lines today is more or less where Heaviside left it. The line could be anything from an iron wire slung on wooden poles with an earth return to a sophisticated suboceanic cable; today it would be a fiber-optic cable. Early telegraphers thought of a line simply as an inert conduit along which a battery pumped a kind of fluid called electricity, but they were puzzled at the way pulses became smeared-out when sent along suboceanic cables, requiring the signaling speed to be slowed so that each pulse could be distinguished from the next. Faraday had explained that this happened because the cable acted as a giant electrical store, or capacitor, and William Thomson had encapsulated this principle in a formula. Thomson had intended his formula to apply only to long suboceanic cables worked at low speed, but telegraphers who didn't understand the mathematics had misapplied it to high-speed land lines, with sometimes bizarre results. Tact was never in Heaviside's repertoire—he mocked the ignorance of the senior post-office engineers unmercifully in his papers and soon became their bête noire. Luckily, he found a friend in C. H. W. Biggs, the editor of the weekly trade journal the *Electrician*. Biggs knew he was running a risk by taking on this enfant terrible. He also knew that scarcely any of his readers could follow Heaviside's increasingly abstruse mathematical papers, but, although no mathematician himself, he sensed that here was something new and important. His support for the maverick writer eventually cost Biggs his job, but meanwhile, Heaviside forged ahead. In his hands, the transmission line became a complex piece of equipment with properties—capacitance and inductance—that corresponded to elasticity and inertia in mechanical systems, and resistance, which was akin to friction. Using these quantities, Heaviside generalized Thomson's result to produce what is still called the "telegrapher's equation."

Articles in the *Electrician* were also an outlet for Oliver's puckish sense of fun. His satirical sallies must have raised a chuckle in the editorial office because Biggs published many passages that more cautious editors would have cut out. The first paragraph of Heaviside's first article must have given Biggs some idea of what he was getting himself into.

> The daily newspapers, as is well-known, usually contain in the autumn time paragraphs and leaders upon marvellous subjects which at other times make way for more pressing matters. The sea serpent is one of these subjects.[3]

Heaviside went on to make sport of a press report about a boy with a stick who could detect water deep underground and only then got down to the business of the paper, which was a masterly account of a strangely neglected topic—the use of the earth as a return conductor in telegraphy. Heaviside loved pricking dignity, and one of his favorite targets was the church: he mocked the pompous solemnity of archbishops and, in one article, mischievously claimed that Ohm's law was evidence of divine creation—God had arranged it to save electricians the trouble of the laborious extra calculations that would have been needed had it not been true.

Cheered by such diversions, Heaviside worked his way, bit by bit, through Maxwell's *Treatise*. At first he was as perplexed as anyone by the great man's theory of electromagnetism, but eventually he managed not only to master the theory but also to re-express it in a form that was much easier to grasp—the form, in fact, in which the theory is generally expressed today. The achievement was later described by another Maxwellian, George Francis Fitzgerald:

> Maxwell, like every other pioneer who does not live to explore the country he opened out, had not had time to investigate the most direct means of access to country nor the most systematic way of exploring it. This has been reserved for Oliver Heaviside to do. Maxwell's *Treatise* is encumbered with the debris of his brilliant lines of assault, of his entrenched camps, of his battles. Oliver Heaviside has cleared these away, has opened up a direct route, has made a broad road, and has explored a considerable trace of country. The maze of symbols, electric and magnetic potential, vector potential, electric force, current, displacement, magnetic force and induction, have been practically reduced to two, electric and magnetic force.[4]

If asked how he did it, he would probably have said "by hard work," and indeed it was. But there were two key components. One was the creation of a language for describing how vectors—quantities that have both magnitude and direction—vary in space. It became accepted as the natural language for the field. Called simply "vector analysis," it fits the part so well that it is hard for today's students to imagine a time when it didn't exist. Each vector in three-dimensional space is represented by a single letter. (Pictorially it could be represented by an arrow of given length and direction.) With the single letters comes an algebra that enables mathematical relationships to be set out in simple, or at least simple-looking, equations that are independent of any coordinate system. The idea came from Maxwell's use, in his *Treatise*, of a form of quaternion representation. As we've seen, quaternions were an elegant but fearsomely complicated creation of the Irish mathematician Sir William Rowan Hamilton. They had, in effect, a vector part and another part that was an ordinary number, or scalar. Heaviside experimented with quaternions but found them largely useless, so he separated the vector and scalar parts and worked out a new algebra for the vectors. He later discovered that, in America, Josiah Willard Gibbs had also found disappointment in quaternions and had independently devised exactly the same vector algebra. Heaviside was happy to share the credit—he was always generous with praise where he thought it was due—and, in any case, it was an honor to share credit with a man like Gibbs.

The second component in Heaviside's simplification of Maxwell's theory was to concentrate on the field forces and push the quantities called potentials into the background. The forces, to him were "real," but the potentials were "metaphysical"[5] and he decided "to murder the whole lot."[6] This way, with a little rearrangement, he was able to reduce Maxwell's twenty equations, or eight in "quaternion" format, to four. Here they are, for a point in empty space where no currents or charges are present:

$$\text{div } \mathbf{E} = 0$$
$$\text{div } \mathbf{H} = 0$$
$$\text{curl } \mathbf{E} = -\mu \partial \mathbf{H}/\partial t$$
$$\text{curl } \mathbf{H} = \varepsilon \partial \mathbf{E}/\partial t$$

$\mathbf{E}$ and $\mathbf{H}$ are the electric and magnetic field forces—the mechanical forces that would be exerted on a unit electric charge or a unit magnetic pole placed at the point. $\partial\mathbf{E}/\partial t$ and $\partial\mathbf{H}/\partial t$ are their rates of change with time, and μ and ε are the fundamental constants of magnetism and electricity.[7] Div, short for divergence, and curl are ways of describing how the vectors vary in a small region surrounding the point and, as we've seen, had already been identified and named by Maxwell. When charges or currents are present, the equations acquire extra symbols to represent charge density and current density, but they are still astonishingly simple.[8] They have been called the *Mona Lisa* of science; it seems a marvel even to professional physicists that they give rise to all the seemingly complicated phenomena of electricity and magnetism. The first two equations imply that electric and magnetic forces obey an inverse-square law, and the third and fourth imply that disturbances in the field will spread out as electromagnetic waves with speed $1/\sqrt{(\mu\varepsilon)}$, which is the speed of light.[9]

Heaviside had given us the four equations that were to become famous. It is right that they are known as Maxwell's equations but they are, in part, Heaviside's creation, too.[10]

With his background in telegraphy, Heaviside was intrigued by the way that electromagnetic energy moved. By Maxwell's theory, energy was *located* in space—at any instant, each part of space contained a definite amount of energy. When a change occurred in the field, some parts of space would gain energy and others lose it, but energy couldn't be simultaneously destroyed at one point in space and created at another, because this would involve actions occurring at a distance—the very concept that Maxwell and Faraday had set out to banish. Energy had to *flow*, and Heaviside worked out how it flowed. The rate of energy flow was equal to the product of the electric and magnetic field forces, its direction was at right angles to both, and it was greatest when the electric and magnetic forces were at right angles to one another. He expressed the law of energy flow compactly in vector form as:

$$\mathbf{W} = \mathbf{E} \times \mathbf{H}$$

where **W** is the energy-flow vector, and **E** × **H** is the vector product[11] of the electric and magnetic field forces.[12]

This was a great result, but soon after it was published in the *Electrician*, Heaviside found that he had been scooped by John Henry Poynting, a former student at the Cavendish who had left Cambridge shortly after Maxwell died and was now professor of physics at Birmingham University—Poynting had published the same result a few months earlier in the Royal Society's journal. Heaviside's response to the disappointment shows two sides of his character: he always generously acknowledged Poynting's priority but never failed to mention his own part in the matter. The Poynting vector, now known to all students of electromagnetism, bears its own aide-memoire, as it *points* in the direction of flow.

Poynting took the glory, but it was Heaviside who did the most to explore the consequences of the new discovery. What he found almost defies belief, even today. In an electrical circuit, no energy passed through the wires themselves—they merely acted as a guide for the flow of energy through the surrounding space. The only energy flow inside the wires was *inward*, and that was just the portion of energy that was dissipated as heat! What about the electric current—didn't that flow in the wires? Yes it did, but its energy was borne by the accompanying fields—the lines of magnetic force that encircled a current-bearing wire and those of electric force that spread out radially from it, like spokes. By the new formula, the direction of energy flow was at right angles to both these fields and so ran parallel to the wire. Very nearly so, anyway: the lines of energy flow near the wire converged ever so slightly and, when they hit the wire, turned sharply inward to be converted into heat.

By 1888, Heaviside had been living much the same life for fourteen years, writing papers that nobody seemed to read and rarely traveling farther from home than his feet could take him. The old feeling of self-sufficiency—that his discoveries were all the "meat and drink" he needed—was wearing thin. He wanted his voice to be heard. Then he happened to read a report of a talk by Oliver Lodge, professor of physics at the University College in Liverpool, and he saw his name mentioned. Referring to electromagnetic waves, Lodge had said:

> I must take this opportunity to remark what a singular insight into the intricacies of the subject, and what a masterly grasp of a most difficult theory are to be found in the eccentric, and sometimes repellent, writings of Mr. Oliver Heaviside.[13]

Repellent was, presumably, a warning that some readers might find Heaviside's gratuitous observations on sundry topics like archbishops to be in poor taste. Anyway, what was *repellent* when set alongside *masterly*?—this was the first public recognition that Heaviside had received in his life, and he was overjoyed. He straightaway wrote to Lodge to ask for a full text of his talk and soon found that he had another admirer, Lodge's friend George Francis Fitzgerald, who was professor of natural and experimental philosophy at Trinity College, Dublin. Like Heaviside, Lodge and Fitzgerald had been captivated by Maxwell's work and both had been trying, first in isolation and then with mutual support, to carry it on. Now Heaviside, the independent recluse, had gained true friendship on his own terms, and the three of them, united in a common cause, became firm friends and formed the core of the group that came to be called the Maxwellians. They were soon to be joined by a fourth person from an unexpected quarter.

Lodge was a clay merchant's son from Staffordshire who hated the trade but endured it until he had a chance to escape when reaching his majority. As a teenager, he had heard John Tyndall speak, and from that moment knew what he wanted to do in life. He worked his way to University College, London, and gained a doctorate before being appointed to the professorship at Liverpool. Forceful and extroverted, Lodge pursued his science with doggedness and passion and had a penchant for mechanical models. We can get an idea of his style from a contemporary's opinion of his book *Modern Views on Electricity*:

> Here is a book intended to expound the modern theories of electricity and to expound a new theory. In it there are nothing but strings which move around pulleys, which roll around drums, which go through pearl beads, which carry weights; and tubes which pump water while others swell and contract; toothed wheels

> which are geared to one another and engage hooks. We thought we were entering the tranquil and neatly ordered abode of reason, but we find ourselves in a factory.[14]

Although he saw the need for mathematics and could, with effort, follow the work of others, Lodge was at his best when experimenting. Fitzgerald, on the other hand, was a gifted mathematician. Born into one of Ireland's patrician Protestant families, he sailed through his degree classes at Trinity College, Dublin, and gained one of the prized fellowships. Everything seemed to come easily to him, and, perhaps for this reason, he lacked the stern mental discipline that comes from sustained toil. Heaviside said of him: "He had, undoubtedly, the quickest and most original brain of anybody."[15] Fitzgerald often failed to follow up his own ideas, claiming in his unpretentious way that he was too lazy, but he dispensed them freely and had an influence on late nineteenth-century physics that extended way beyond his own published work. He commanded tremendous respect, and it was chiefly through his influence that other British physicists began to pay serious attention to Maxwell's theory and to Heaviside's work. Fitzgerald and Lodge met in 1878 at a conference in Dublin and quickly struck up a friendship. Both had become fascinated by Maxwell's writings and each had been trying, in his own way, to take the work forward.

Now they could share ideas, and Lodge set himself the goal of producing and detecting electromagnetic waves. According to Maxwell, they would be produced whenever an electric current changed; the problem lay in detecting them. But light waves, Lodge reasoned, were easy to detect, so why shouldn't he tackle the problem from the other end and try to produce light by electromagnetic means? He tried various methods—for example, passing a current across the contact point between a rapidly spinning carbon disc and another piece of carbon held against it—but failed utterly; he didn't get anywhere near the required frequency. Meanwhile, Fitzgerald had worked out theoretically that the amount of energy radiated by a pulsating electrical circuit was proportional to the *fourth power* of the frequency. This meant that at low frequencies, say several hundred cycles per second, the radiated energy would be weak, but that at frequencies of several million cycles per second, it should be strong enough to be detectable. And the wavelength, the distance from peak to peak of the wave, would then be a few meters, short enough to be measured in a laboratory. Moreover, Fitzgerald believed, the means of achieving such high frequencies already existed—all one had to do was to discharge a Leyden jar through a suitable circuit. One still had to detect the waves, but Fitzgerald had two good ideas here. One was to reflect a wave back toward its source and so form a *standing wave* (a wave that doesn't travel but just vibrates up and down in one place) which would be much easier to detect. The other was to use a detector circuit that was *tuned* to the frequency of the wave. These two elements did indeed prove crucial, but a third was lacking. As Fitzgerald put it, "the great difficulty is something to *feel* these rapidly alternating currents with."[16] No known instrument seemed to be up to the job, but, as we will see, somebody did find a simple and effective way to "feel" the currents. And it had been available all along.

One might have expected practical-minded Lodge to take up the challenge, but for a while he was too busy with lectures and other work to carry out serious laboratory experiments. Then he accepted an invitation from the Society of Arts to give a series of talks on lightning protection, and, in preparation, he tried a few quick experiments with discharging Leyden jars—the sparking from the discharge, he thought, would simulate lightning. When Lodge discharged his jar, sparks appeared between the ends of wires connected to it. This was expected, but he found that the sparking could be made weaker or stronger by varying the lengths of the wires. This was interesting, but only when prompted by a junior colleague did Lodge realize that he had stumbled on evidence of Maxwell's electromagnetic waves. Waves streaming from the discharge through the space alongside the wires had been reflected from the wire ends, and what he had detected were the resulting standing waves—the stationary waves that vibrate in one place and occur whenever a traveling wave combines with its own reflection. He had discovered electromagnetic waves along wires. In his second talk to the Society of Arts in February 1888, Lodge presented a brief sketch of his new evidence supporting Maxwell's theory. He knew that more rigorous experiments would be needed to get the result fully recognized, but there was plenty of time to do this before the big September meeting of the British Association for the Advancement of Science in Bath.

Preparations done, Lodge happily set off for a summer holiday walking in the Alps. For the journey he took some journals he hadn't had time to read. As the train pulled out of Liverpool, he turned to the July issue of the German journal *Annalen der Physik und Chemie* and was astonished to find that Dr. Heinrich Hertz of the Technische Hochschule at Karlsruhe had already produced and detected electromagnetic waves not only along wires but *in free space*. Moreover, he had measured the speed of the waves and had shown that they could be reflected, refracted, and polarized, just like light. Lodge was devastated—his own efforts seemed puny in comparison—but his disappointment was soon overtaken by admiration of Hertz's work and genuine pleasure at the results. His own findings would now play only a small part at the British Association's September meeting in Bath, but there was a much bigger story to tell. Hertz's results had provided clear proof of Maxwell's theory of the electromagnetic field and had finally laid action at a distance to rest. Still, many of the delegates at Bath would be asking: who is Heinrich Hertz?

Hertz grew up in a comfortable home in Hamburg. His father, a barrister, was from a long line of Jewish merchants but had converted to Christianity, and his mother came from generations of Lutheran preachers in the south of Germany. With this eclectic background, the boy developed wide interests from an early age, and shone at everything: languages, classics, mathematics, and sports. Faced with a choice of "klassiker" or "techniker" classes, he managed to combine the two by changing schools several times and, at one stage, studying at home. He enjoyed practical work and, on leaving school, studied engineering at Dresden and Munich before realizing that his true vocation lay in mathematical and experimental physics. There was then only one place to go, and, at twenty-one, he moved to Berlin University, where he became the star pupil and then the assistant of Hermann von Helmholtz—the country's most celebrated scientist.[17]

Helmholtz had a vast range of interests, but the topic that held most of his attention at this time was electromagnetism, and he was one of the few top-ranking physicists to take Maxwell's theory seriously. In Helmholtz's view, three theories were more or less equal contenders: those of his compatriots Wilhelm Weber and Franz Ernst Neumann, which were both based on action at a distance, and Maxwell's. It was important, he thought, to establish by experiment which was correct. Prompted by Helmholtz, Hertz tried an experiment to detect displacement currents, but he found nothing. The currents, if any, were too weak to register on the most sensitive instruments available. Still, the work honed his experimental skill, so when luck gave him the slenderest opportunity a few years later, he was ready to exploit it. Meanwhile, he needed to gain teaching experience and so for two years worked as an unpaid lecturer at Kiel University. In his free time, he turned to the theoretical aspects of Maxwell's theory. Amazingly, he arrived at the same equations as Heaviside, though in the old triple-equation format. When the two later came to know one another, Hertz graciously acknowledged Heaviside's priority and told him he believed:

> . . . you have gone further than Maxwell and that if he had lived he would have acknowledged the superiority of your methods.[18]

In 1885, Hertz gained the post of professor of experimental physics at the Hochschule in Karlsruhe. Within a year, he had married the daughter of one of the other lecturers and was hard at work in the well-equipped laboratory, trying again to find the faintest sign of displacement currents in insulators. Attempt after attempt yielded no result. In the end he succeeded, but in the course of these attempts he found a far more effective way to verify Maxwell's theory.

Among the stock apparatus was a pair of so-called Knochenhauer spirals, flat coils insulated by sealing wax, that were intended to give a graphic demonstration of Faraday's principle of induction. A spark-generating circuit induced sparking across the terminals of a another circuit that was separate from the first but magnetically linked to it across a small air gap. One day, possibly while setting up a demonstration for a class, Hertz was surprised to see sparks also coming from a stray wire alongside. What was going on? At this stage, he didn't really know what he was looking for, but he followed his intuition and felt his way toward a great discovery. A year before Lodge, Hertz discovered electromagnetic waves along wires. Both had taken advantage of chance observations, but Hertz found something Lodge had missed—something to "feel" the waves with—and it was nothing more than a loop of wire with a small gap between its ends across which sparks could jump. If the detector loop was the right size and shape, it would be tuned to the frequency of the waves; they would set it resonating and generate enough electromotive force in the wire to make sparks jump across the gap. It sounds simple, but even for Hertz, the most gifted of experimenters, results came only after many hours of trial and error and were the fruit of determination as much as skill.

Waves along wires were exciting, but the ultimate test of Maxwell's theory would be to detect waves in space. Hertz's primary sparking circuit became his transmitter, and the loop of wire with the spark gap his detector.

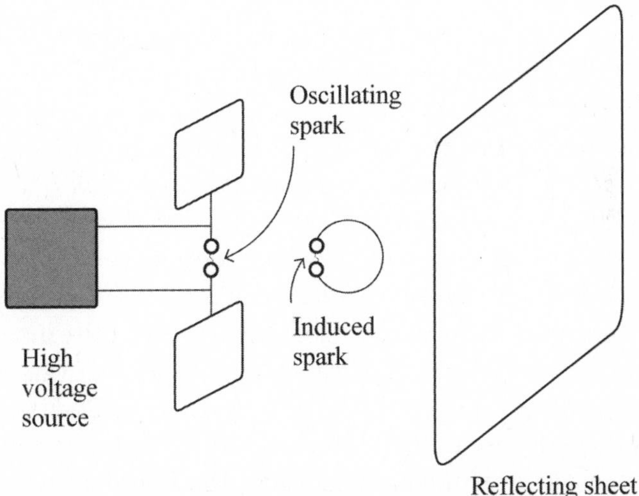

Fig. 16.1. Schematic layout of Hertz's apparatus for producing and detecting electromagnetic waves. (Used with permission from Lee Bartrop.)

He tried many variants of both and, with the spark terminals in the detector set as close together as possible, he carried it around the room looking for sparks. Sure enough, faint sparks appeared. We now come to the historic scene that introduces this book. Using a large zinc sheet as a reflector, he moved the detector around and found some places where there was no sparking and others where the sparking was strongest. This was evidence of a standing wave that can only have been formed by a traveling wave combining with its reflection from the zinc screen. He had produced and detected electromagnetic waves in space.

For some this might have been enough, but for Hertz it was just the beginning. In a brilliant series of experiments, he examined every aspect of the new waves. He found that although the waves were reflected by any metal surface, they passed unimpeded through thick wooden doors. He also showed that the waves traveled at the speed of light; that they could be polarized, just like light; and that they could be refracted in the same way that light is through a glass lens or a prism. In a sense, his waves *were* light waves, but of much longer wavelength than visible light; we use them as radio waves today.

In 1888, Hertz presented his findings in a magnificent series of experimental papers—it was the second of these that Lodge had read on the train out of Liverpool. They were written in a matter-of-fact style, with no grand announcement, and at first attracted very little attention. Given Helmholtz's apparent enthusiasm for Maxwell's theory, one might have expected some excitement in Germany, but physicists there had been raised on the action-at-a-distance theories of Weber and Neumann, and even Helmholtz had interpreted Maxwell's theory in a way that held onto some of the action-at-a-distance notions and was, according to the British Maxwellians, plain wrong. It failed to explain the flow of electromagnetic energy in space, and without energy flow there could be no waves, no matter how you jiggled the mathematics. Heaviside put it bluntly, as usual:

> Helmholtz's theory seems to me as if he had read Maxwell all at once, then gone to bed and had a bad dream about it, and then put it down on paper independently; his theory being Maxwell's run mad.[19]

Hertz, though, revered Helmholtz and didn't realize at first what a catastrophic blow he had dealt to his mentor's ideas. He knew his discoveries were significant, but he seemed to surpass even Maxwell in modesty and said he was content to let others judge their worth. Across the English Channel there was no such restraint. The British Maxwellians had made the theory their own and had no doubt of its validity. All they lacked was clear physical proof to convince skeptics, and this Hertz had handsomely supplied. He was a hero. They showered him with praise, welcomed him into their ranks with joy, and set about promoting his work with tremendous enthusiasm. Fitzgerald told the assembled company at the British Association's big meeting in Bath of the "beautiful device" by which Hertz had succeeded; Lodge made a replica of Hertz's apparatus, which he demonstrated at every opportunity; and Heaviside wrote to thank Hertz for killing off the action-at-a-distance theories:

> I recognized that these theories were nowhere, in the presence of Maxwell's, and that he was a heaven-born genius. But so long as a strict experimental proof was wanting so long would these speculations continue to flourish. You have given them a death blow.[20]

It was later often said, with no more than slight exaggeration, that news of Hertz's discoveries reached Germany by way of England. In 1890, the Royal Society awarded Hertz its Rumford Medal, and he came to London to collect it. One free evening, Hertz, Lodge, and Fitzgerald dined together at the Langham Hotel. They must have felt the presence of a metaphorical empty chair. Heaviside, who habitually turned down invitations, would surely for once have left his room and joined them, but he was by now living two hundred miles away in Torquay. The four had become a close and mutually supportive group—four very different individuals joined in a common cause. They had brought Maxwell's theory—in truth both Maxwell and Faraday's theory—to the world.

CHAPTER SEVENTEEN

A NEW EPOCH

1890 ONWARD

The epoch that, in Einstein's words, began with James Clerk Maxwell was under way. The Maxwellians had opened up huge opportunities, but there were still skirmishes to be fought and shocks to be borne.

When Heinrich Hertz visited London in 1890, he had been only narrowly dissuaded from making the 400-mile round journey to Torquay to see the reclusive Heaviside. They had built up a firm friendship through letters, but neither knew this would be their only chance to meet. Hertz died from a rare bone disease in 1894 at the tragically early age of thirty-six. We can never know what else he would have achieved, but the other Maxwellians still had parts to play.

Oliver Heaviside was always in a fight with someone or other, and in the early 1890s his chief adversary was Maxwell's old friend P. G. Tait. They clashed over the new system of vector analysis, by which Maxwell's theory had been summarized in the four now-famous equations. In Tait's view, Heaviside (and, independently, Gibbs in America) had committed sacrilege by mutilating William Hamilton's beautiful quaternions. Heaviside already had widespread opposition: some from people who couldn't make heads or tails of either quaternions or the new vector notation, and some from others like William Thomson who could but preferred to stick to the old triple-equation method with its so-called Cartesian (x, y, z) coordinates. However, Tait was his fiercest opponent and both enjoyed a good scrap. When Tait described vector analysis as "a hermaphrodite monster," Heaviside responded by calling Tait a "consummately profound metaphysicomathematician" and added:

> "Quaternion" was, I think, defined by an American schoolgirl to be "an ancient religious ceremony." This was, however, a complete mistake. The ancients—unlike Prof. Tait—knew not, and did not worship Quaternions.[1]

Exchanges like these appeared in the scientific journals and kept readers entertained for more than a year. As usual, Heaviside was right: today, his vectors are everywhere and quaternions are hardly to be seen.

Meanwhile, Oliver Lodge was trying to take Hertz's experimental work further, using an ingenious new receiving device called a coherer.[2] He made good progress but was once again overtaken, this time by a young Italian. Guglielmo Marconi was having similar success with similar equipment on the family estate near Bologna; he had seen very early the huge commercial potential of radio telegraphy and was set on making his fortune. Having failed to get sponsorship in his own country, he came to England and persuaded the Post Office's senior engineer, William Preece, to back him. Things went well, but the astute Marconi soon felt the bonds of state patronage tightening, cut himself free, and started his own company with help from an English cousin who had influence in the money markets of the city of London. Not only a resourceful inventor but also a persuasive salesman, Marconi was soon running well-publicized ship-to-shore trials in the Solent, a strait separating the Isle of Wight from the English mainland. He knew very little theory and worked mostly by trial and error, but now he could afford to employ the best engineers and hired Maxwell's old student, Ambrose Fleming.

With Fleming's help, Marconi planned to stage the most dramatic event possible by sending a wireless telegraph message across the Atlantic Ocean. Most physicists thought the venture was doomed because the transmitted waves wouldn't follow the curve of Earth's surface but would go straight off into space. However, Marconi's intuition said otherwise: the project went ahead, and in 1901 a signal from Poldhu in Cornwall was received by a kite-borne antenna in Newfoundland. Men had begun to harness Maxwell's electromagnetic waves for their own use, and it was as though Earth had shrunk.

Wireless telegraph became standard equipment on ships, sound radio followed, then television, radar, cell phones, and worldwide transmission of signals via satellites. The miracle of near-instant communication without wires over great distances became taken for granted—a part of everyday life.[3]

Another aspect of the new technological epoch was the growing use of electrical energy in homes, in factories, and in transportation. It stemmed from Faraday's discoveries of the electric motor in 1821 and electromagnetic induction in 1831, but things only really got going much later in the century when, thanks to Swan's and Edison's filament light bulbs, domestic electric lighting became a commercial proposition. There was then a need to generate and distribute electricity widely and efficiently. When this was met—first by using direct-current methods and then by Nikola Tesla's brilliant multiphase alternating-current generators and distribution systems—the way was open for the development of all kinds of electrical machinery for industrial and domestic use, and for transport. Every generator, every motor, and every transformer that we use today depends on the interaction of electric and magnetic field forces. In short, not only our communications but also almost our whole way of life has come to depend on technology that exploits the electromagnetic field—a feature of the physical world that was undreamed of until it was first envisaged by Faraday, then elucidated by Maxwell.

But even more significant than the advance in technology is the way that Faraday and Maxwell's concept of the electromagnetic field transformed scientists' view of the physical world. During the late decades of the nineteenth century a sea change was gradually taking place within the physics community as more and more people grasped the truth of Maxwell's warning: mechanical models cannot be relied on to explain physical phenomena, and to use them risks confusing representation and reality.

Faraday and Maxwell's field was intangible, and space was not just an empty geometrical container for bodies with mass but a coherent interconnected system bearing the energy of motion. It was the seat of action, rather than just an empty backdrop for Newton's point particles being propelled by straight-line forces. These were ultraradical concepts for nineteenth-century minds trained to think only of things that could be touched and measured. Properties of the field, like the electric and magnetic intensities, were abstract quantities—all they had in common with the quantities in Newton's laws of matter in motion was that they obeyed dynamical equations. Maxwell had replaced a universe in which tangible objects interacted with one another at a distance by one where abstract fields extended throughout space and interacted only *locally* with tangible objects.

His equations of the electromagnetic field came to take on a life of their own, divested of any mechanical association—the abstract mathematical language of the field was sufficient in itself. While Fitzgerald, Lodge, and others were searching for ingenious quasimechanical ways to interpret Maxwell's theory, Heinrich Hertz gave the simplest and best explanation. He said, "I know of no shorter or more definite answer than the following—Maxwell's theory is Maxwell's system of equations."[4]

As Freeman Dyson has aptly observed, Maxwell's theory becomes elegant and clear only after one has given up the need for a mechanical model.

Here was a *physical* theory expressed solely by its equations, and it brought about a profound change in physicists' concept of reality. The startlingly new idea was that reality exists *on two levels*. Beneath all the things that we can touch and feel (or make models of) lies a deeper reality that expresses itself in the language of mathematics. In this underlying layer are quantities, like the electric and magnetic field intensities, that are quite different from anything we can access with our senses. They must, in some sense, be real, because they give rise to all the mechanical forces that we *can* feel and represent in models, but the only way we can describe them is through abstract symbols in equations.

Although the idea of a two-tier reality was implicit in Maxwell's theory, he never articulated it in quite this way and it didn't fully take hold until well into the twentieth century. Nevertheless, the ground was shifting. Newton's laws of matter and motion, for two centuries the bedrock of natural philosophy, no longer provided a sufficient base for scientific thinking about all phenomena in the physical world. As Einstein put it:

> Since Maxwell's time, physical reality has been thought of as represented by continuous fields and not capable of any mechanical interpretation. This change in the conception of reality is the most profound and fruitful that physics has experienced since the time of Newton.[5]

The theory of the electromagnetic field, encapsulated in four equations, had ushered in a new epoch and was gradually gaining general acceptance but was itself soon jolted by two discoveries made around the turn of the century. The first was the electron.

Faraday and Maxwell believed in the primacy of the field. Their electric and magnetic fields filled all space, including those parts occupied by matter. Matter interacted with the field by modifying the properties of the parts of the field with which it shared space, and one effect of this was the appearance of electric charge on the surfaces of conducting substances immersed in an electric field. In an insulating medium such as glass or air, the field took the form of an electric displacement, or polarization of the insulating particles along the lines of force; and where the lines terminated, say on two conducting metal plates, the surfaces of the plates, by virtue of their contact with the field, *appeared to contain* a charge, positive on one plate and negative on the other. Electric currents were similarly effects of the field. Taking these thoughts further, Heaviside had explained that current-carrying wires were merely guides for the passage of energy through the surrounding space. All this was in complete contrast to the earlier view that electricity was some kind of fluid substance that existed in conductors, and that electric and magnetic forces resulted from the charges or currents in conductors acting on one another at a distance. The field view had triumphed when Hertz found the electromagnetic waves that Maxwell had predicted, but not all theorists were content to believe that electric charge was simply an artifact of the field.

Among the doubters were the great Dutch physicist Hendrik Antoon Lorentz and Joseph Larmor, an Ulsterman who had happily settled into the life of a don at Cambridge. They began to construct versions of the field that allowed electrically charged particles to exist in their own right, and, in the end, succeeded in modifying and extending Maxwell's theory. This wasn't all. In 1897, Lord Rayleigh's successor at the Cavendish, J. J. Thomson, discovered the electron, a material particle with an intrinsic electric charge. Here was empirical evidence of the theoretical particles that Lorentz and Larmor had postulated. They had, in a sense, completed Faraday's and Maxwell's great project by establishing the true relationship between the field and matter: electric charges reside in matter, but their effects are transmitted by the surrounding field, with which their own fields are intertwined. In his "dynamical" aether, Maxwell had retained the quasimechanical concepts of flywheel-like kinetic energy and spring-like potential energy, but Lorentz replaced it with a purely electromagnetic aether that interacted with electrons whose energy of motion was entirely due to their own electromagnetic fields. The mechanical school of thought was relinquishing its last hold: scientists had begun to see electromagnetism not merely as a companion to the traditional laws governing the motion of matter but as the new conceptual foundation for all of their studies of the physical world.

The second of the great discoveries at the turn of the century came like an earthquake. Maxwell's theory of electromagnetism predicted that so-called black bodies should radiate much more energy at high frequencies than they actually did. (A black body is a theoretical object that absorbs all the radiation that hits it.) Nature seemed to have some hidden mechanism that chopped off the high-frequency end of the radiation spectrum, but no one had any idea what it was. Then, in 1900, Max Planck, professor of theoretical physics at Berlin University, found the solution. In what he called "an act of desperation,"[6] he cobbled together a formula that matched experimental results very well, but with a proviso: the energy radiated or absorbed by black bodies had to occur only in discrete amounts, or "quanta" that were proportional to the frequency of the radiation. For a while, Planck and others mistrusted his monstrous creation and went on looking for a more plausible alternative, but in 1905, a junior clerk in the Bern patent office named Albert Einstein boldly took things even further by proposing that Planck's quanta were not merely specified amounts of radiation but indivisible, discrete *packets* of radiation, now called *photons*.

Amazingly, Einstein produced readymade empirical evidence to support his claim: it explained the hitherto-inexplicable photoelectric effect. Experimenters had found that a beam of ultraviolet light could dislodge electrons from the surface of a metal object, causing them to be emitted. The shorter the wavelength of the light, the higher was the energy of the electrons, but, strangely, when the intensity, or "brightness," of the beam was weakened, the energy of the electrons stayed the same, though their number decreased. Maxwell's theory on its own couldn't explain this, but Einstein's photons did so perfectly. Each photon was a separate packet containing an amount of energy that depended only on the wavelength of the light (or its frequency), so when the beam was weakened, it contained fewer photons, but each one had the same energy as before. Consequently, fewer electrons were emitted but the energy of each one was the same, identical to that of the photon that had dislodged it. The electron's energy was equal to $h\nu$, where ν was the frequency of the electromagnetic wave and h was a new quantity that came to be called Planck's constant. Its value, $6.62606957 \times 10^{-34}$ joule-seconds, demonstrates the minute size of energy quanta.

So, the electromagnetic waves that Faraday had envisaged in his "Ray-vibrations" talk, that Maxwell had predicted mathematically in his "Dynamical Theory" paper, and that Hertz had produced and detected in his laboratory not only had the properties of a continuous wave but also, in an apparent paradox, also behaved like discrete packets or particles. How could the great theory of the electromagnetic field be reconciled with this apparently shattering revelation? In the course of the quantum's assimilation into mainstream physics in the early decades of the twentieth century, this wave-particle duality became a tenet of a new theory. Niels Bohr, Werner Heisenberg, Erwin Schrödinger, and Wolfgang Pauli led the way in the creation of quantum mechanics. They and others, notably Paul Dirac, also eventually adapted the "classical" field concept of electromagnetism for minute-length scales (those on the order of Planck's constant) by "quantizing" the field. Hence have come all the great field theories of modern physics, such as quantum electrodynamics and the Standard Model—the reigning model of intercoupled particles and fields by which today's physicists are carrying on Faraday's quest to unify all known forces.

There's an interesting sidelight on these great developments. The creators of quantum electrodynamics didn't just use Heaviside's compact four-equation version of Maxwell's theory. As one of them, Richard P. Feynman, explained: "In the general theory of quantum electrodynamics, one takes the vector and scalar potentials as the fundamental quantities."[7]

These "fundamental quantities" are the very ones that Heaviside had eliminated when condensing Maxwell's equations, so the wisdom of Maxwell in keeping options open by retaining all the equations has been handsomely borne out.

Our story concludes with yet another great discovery—one in which the "classical" theory of the electromagnetic field played a central role—and it starts with what had been the last bastion of the old mechanical school, the aether. Even though Maxwell had abandoned his mechanical model, he was ambivalent about the aether and was never able to do away with the notion completely. In his "Dynamical Theory" paper, he still required a medium, or aether, even if it was, in effect, just a set of properties without a specified mechanism. Others persisted in constructing ever more elaborate mechanical models of the aether: Lodge's favorite model had a rack and pinion mechanism; Fitzgerald's had pulleys and bands; and William Thomson proposed one, called the vortex sponge, with a mechanism so unusual that even he couldn't find exact equations to describe it.

Since the aether occupied all space, Earth must move through it, as a ship moves through the sea. This motion was called the "aether drift," and physicists set out to measure it. The measurement required such precision that Maxwell himself doubted whether it could be achieved in any laboratory; he proposed instead a method using observations of Jupiter's moons. Nothing came of that, but a young American, Albert Michelson, took Maxwell's doubts about earthbound methods as a challenge and developed his interferometer, an instrument that used the tiny wavelengths of light as measurement units and so made possible a degree of precision hitherto undreamed of. With his colleague Edward Morley, he devised an experiment to measure the difference in speed of the two parts of a light beam that had been split at right angles.[8] Michelson and Morley carried out their experiment in Cleveland, Ohio, in 1887. Even the smallest difference in speeds would have established for all time that the aether did indeed exist, but to the experimenters' consternation, the speed of light in both directions was identical, and repeated trials gave the same result. This was a huge disappointment, and at first the experiment was seen as no more than another failed attempt to measure the aether drift. Michelson himself rarely spoke of his result and never realized its significance.

Heaviside was, at about the same time, investigating the implications of Maxwell's theory for the behavior of moving electric charges, and in a paper published in 1889 he made the extraordinary claim that the field of a point charge moving with velocity v relative to the aether would contract in its direction of motion by the factor $\sqrt{(1 - v^2/c^2)}$, where c was the speed of light: If the charge reached the speed of light, its field would be squashed flat. This was the first appearance of the factor $\sqrt{(1 - v^2/c^2)}$, which has become very familiar to physicists and is sometimes called simply the relativistic factor. Heaviside's friend Fitzgerald took the idea further and proposed that all matter behaved in this way. If every object moving relative to the aether contracted by the same factor in its direction of motion, then Michelson and Morley's result was explained—their instruments would contract by exactly the amount needed to compensate for the aether drift, hence the null result. The idea seemed crazy, but Fitzgerald wasn't the only one to think along these lines—Lorentz independently made the same proposal, and the phenomenon came to be called the Lorentz-Fitzgerald contraction. This wasn't all: Lorentz went further by asserting that clocks would slow down by the same factor if they approached the speed of light—if a clock actually reached the speed of light, it would stop.

How could measuring rods shrink and clocks slow? Some began to question the existence of any absolute measures of space and time. One was the great French mathematician and occasional physicist Henri Poincaré. In a book published in 1902, he wrote:

> There is no absolute uniform motion. No physical experience can therefore detect any inertial motion. There is no absolute time. Saying that two events have the same duration is conventional, just as saying they are simultaneous is purely conventional, if they occur in different places.[9]

Poincaré was anticipating what is now called the special theory of relativity. In an earlier paper, he had shown that the equivalent mass, m, of a quantity of electromagnetic radiation with energy E is given by an equation with a familiar ring, $m = E/c^2$, but he didn't put everything tightly together and was overtaken by another, just as Lodge had been by Hertz and Marconi.[10] The scene was set for Albert Einstein, who produced his famous paper on special relativity in 1905 within a few months of the one in which he predicted the photon.

Like Heaviside, Fitzgerald, Lorentz, and Poincaré, Einstein studied Maxwell's theory and thought long and hard about its consequences, especially those for time and space. Eventually he found a devastatingly simple and direct approach that the others had missed. The laws of physics, he claimed, must be the same for all observers traveling at a uniform velocity relative to one another. Maxwell's equations were among these laws, and they gave a single value for the speed of light in a vacuum (or in air) irrespective of the observer's motion. In Einstein's view, this was enough on its own to explain Michelson and Morley's result—the speed of light was always the same for any observer traveling at a steady velocity—but what followed from this simple-sounding proposition was one of the greatest shocks ever to hit the world of science. The laws of physics were, indeed, the same for all observers in uniform relative motion. What was *not* the same were their measurements of time and space—any two observers who were moving relative to one another measured both time and space differently. To reconcile their observations required a mathematical transformation of coordinates using the factor $\sqrt{(1 - v^2/c^2)}$; Einstein called it the Lorentz transformation. Using this transformation in conjunction with Maxwell's equations of the electromagnetic field, Einstein calculated that when a body absorbs a given amount of energy from radiation, its inertial mass increases by that amount divided by c^2. What then followed after a few lines of algebra was the famous equation:

$$E = mc^2$$

where E is the intrinsic energy of a body, m is its mass at rest, and c is the speed of light, about 300 million kilometers per second. By Einstein's reasoning, this relationship between mass and energy was a necessary consequence of Maxwell's theory of electromagnetism. It was a finding of tremendous importance to physicists, but nobody at the time dreamed that it would be possible to make a bomb by annihilating a little mass and so liberating a vast amount of energy. As we've seen, this equation had already been published by Poincaré. All the other formulas of special relativity had also been published in one form or another, but it was Einstein who brought everything together in 1905 with a crystal-clear and utterly fresh vision.[11]

Another consequence of Einstein's theory was that nothing could travel faster than the speed of light. In fact, no object with mass could even reach that speed because to do so would require an infinite amount of energy. Remarkably, nature had a speed limit that was completely determined by Maxwell's theory of the electromagnetic field and depended only on the elementary properties of electricity and magnetism.

What of the aether? It needed to operate in a single universal frame of reference of absolute space and time, and Einstein had demolished those by showing that observers who were moving relative to each other measured distance and time differently. So the aether no longer had a home. Nor did it have a reason for existence. No longer was space simply the theater in which the laws of physics performed; coupled with time, it was part of the action. Space and time were entities in their own right; they obeyed the laws of special relativity and, by the same token, had exactly the properties necessary to support the electromagnetic field. As Lorentz went on to show, Maxwell's equations actually *required* space and time to behave as Einstein proposed. So all the efforts of the aether-model builders came to nothing in the end. Or did they? The keystone of Maxwell's theory, the displacement current, had its origin in the idea that the spinning cells in his now-discarded model could be springy. And although all the mechanical models proposed by Lodge, Fitzgerald, Thomson, and others seem bizarre today, they served in their time as stimuli for thought and so contributed to the general development of physical science. They had served as the scaffolding upon which the field theory of electromagnetism was built, to be then kicked away so that the theory could stand on its own, tall and free.

It is often said that Faraday and Maxwell provided the bridge between Newton and Einstein. While true, this statement is incomplete. Newton was known to have attributed his achievements to "standing on the shoulders of giants," and when Einstein visited Britain it was natural for the press to ask him if he had stood on the shoulders of Newton. Einstein replied: "That statement is not quite right; I stood on Maxwell's shoulders."[12] Maxwell would have pointed out that he, in turn, had Faraday's shoulders to stand on. Their partnership made a contribution to physical science, indeed to human knowledge, comparable with those of Newton and Einstein.

Einstein said a new epoch began with James Clerk Maxwell. Maxwell himself would probably have said that it began in 1821 when Michael Faraday first imagined a circular force around a current-carrying wire. Together, they gave future generations a model for the interplay of experiment and theory, where each illuminates a path for the other. Neither man was confined to the role commonly assigned to him by casual historians. Faraday, the renowned experimenter, put forward some of the most imaginative and daring theoretical ideas; and Maxwell, the cerebral theoretician, carried out some of the most demanding experiments. Both knew that no theory counted a jot unless it stood up to the scrutiny of experiment. The dialogue between experiment and theory that they conducted was one of the most fertile ever to occur in science, and it set a priceless precedent for twentieth-century physics.

They also gave us the means to leave behind a mechanical world of rigid bodies and instantaneous straight-line forces operating at a distance, and to move to four-dimensional space-time, where time, length, and mass depend on the observer. Who could have guessed that the Newtonian outlook would turn out to be so parochial, or that there was a new world lying beneath the surface of our everyday reality? The notion of the field has been the portal to the great discoveries of modern physics, leading us to profound questions about the ultimate nature of the universe at scales of huge energy and infinitesimally small length that even far-seeing Maxwell could never have imagined.

Their paradigm-changing discoveries opened the way to today's great research in elementary particle physics, exemplified by the quest for the Higgs field, a field that endows matter with mass and gives it structure. In tacit tribute to Faraday's prescience in seeking to unify nature's forces, and growing from the seeds he sowed using his simple laboratory instruments, physicists today are still trying to unite the fundamental forces of the universe—the electromagnetic force, the weak and strong nuclear forces, and gravity—in a single, unified theory.[13] Their quest requires millions of subatomic collisions to take place in massive particle accelerators in order to reach the enormous energies and minute dimensions where the four forces might be shown to be different aspects of a single, unifying force. The Higgs field has been such an alluring goal that governments have poured billions of dollars and trillions of volts of electricity into machines designed to wrest it into observable reality, and scientists at the Large Hadron Collider at the European Organization for Nuclear Research (CERN) are now celebrating success. Work goes on, and doubtless more discoveries will bring yet deeper questions.

Faraday and Maxwell remind us of what it means to be a true scientist, whose work embodies the ideal of the human intellect trying to understand nature. They were seekers of truth—inquisitive, objective, tenacious, and ethical, and without vanity or worldly ambition. Their generosity of spirit and their humility enhanced their stature as scientists. One might say that, as Victorian gentlemen, it was easier to for them to embody these ideals than it is for anybody today—the business of science was simpler then, and gentlemanliness mattered more—but the characters of Faraday and Maxwell would have shone through no matter what the age, and their greatness encompasses not only their discoveries but also their characters as scientists and as men. If there is something like heroism in science, they are heroes.

The influence of Faraday and Maxwell's work spreads far beyond their achievement in elucidating and unifying electricity and magnetism. Their concept of the field challenged a paradigm that had seemed immutable and, through their powers of thought and experiment, they began to reveal some of nature's deepest secrets. Their theory underlies all the great triumphs of twentieth-century physics, from special relativity to the Standard Model, and has made possible a vast array of new technologies that have transformed the way we live. Their brilliance still inspires our search for scientific truth, and their deep humanity still offers a shining example of how to live a scientific life. Let us hope that a true understanding of these two men and how their theory evolved will illuminate the path to further discovery in years to come.

NOTES

One purpose of these notes is to give sources. A second is to fill in some technical detail where it may be helpful. A third is to shed interesting sidelights, and we hope all readers will enjoy browsing through these pages.

ABBREVIATIONS FOR SOURCES

Bence Jones	Bence Jones, Henry. *The Life and Letters of Faraday*. 2 vols. Philadelphia: Lippincott, 1870.
Campbell and Garnett	Campbell, Lewis, and William Garnett. *The Life of James Clerk Maxwell*. 1st ed. London: Macmillan, 1882. Page numbers refer to the Sonnet Software online version of this book (second edition, 1999).
Hamilton	Hamilton, James. *Faraday: The Life*. London: Harper Collins, 2002.
Harman (1990–2002)	Harman, Peter M., ed. *The Scientific Papers and Letters of James Clerk Maxwell*. 3 vols. Cambridge: Cambridge University Press, 1990–2002.
James (1991–2011)	James, Frank A. J. L., ed. *The Correspondence of Michael Faraday*. 6 vols. London: Institution of Engineering and Technology, 1991–2011.
Researches	Faraday, Michael. *Experimental Researches in Electricity*. New York: Dover Publications, 1965.
Simpson	Simpson, Thomas K. *Maxwell on the Electromagnetic Field*. New Brunswick, NJ: Rutgers University Press, 1997.
Thompson	Thompson, Sylvanus P. *Michael Faraday: His Life and Work*. London: Cassell, 1901.
Williams (1965)	Williams, L. Pearce, *Michael Faraday: A Biography*. New York: Basic Books, 1965.

INTRODUCTION

1. The Wheatstone bridge, a type of circuit used for measuring electrical resistance, has been named after Charles Wheatstone but was actually invented by a colleague, Samuel Christie. However, Wheatstone himself was a prolific inventor, and any spurious credit he gained from the bridge was offset when one of his many true inventions, a system for coding messages, became known as the Playfair cipher, after Lyon Playfair, a chemistry professor who took up politics, became a baron, and vigorously promoted the cipher for official use.

CHAPTER ONE: THE APPRENTICE

1. The epitaph is on Sandeman's tomb in Danbury, Connecticut. He emigrated to America in the 1760s.

2. The description of Faraday's "brown curls" is quoted by Thompson, page 4.

3. Faraday recollected his childhood belief in the *Arabian Nights* in an 1858 letter to Auguste de la Rive, given in Williams (1965), page 552.

4. This instruction by Watts can be found in *The Improvement of the Mind, Also His Posthumous Works*, page 44.

5. This instruction can also be found in *The Improvement of the Mind, Also His Posthumous Works*, on page 33.

6. Faraday described his electrical experiment with a solution of Epsom salts in a long letter to Benjamin Abbott on July 1812, given in Bence Jones, vol. 1, pages 16–22.

7. The remark about Davy's eyes is reported in John Davy's book *Memoirs of the Life of Sir Humphry Davy*, page 136.

8. The "no answer required" incident is reported in James (1991–2001), vol. 1, page xxx.

9. This despondent passage is from Faraday's October 18, 1812, letter to his friend John Huxtable, given in Bence Jones, vol. 1, pages 44–46.

10. Faraday recollected Davy's warning that science was a harsh mistress in an 1829 letter to Davy's biographer John Ayrton Paris. The complete letter can be found in Thompson, page. 10.

11. Instituted by Napoleon and formally designated the Volta Prize, this was an award for research in electrochemistry and had a cash value of 3,000 francs.

12. Davy's letter about the Napoleon Prize is quoted by John Ayrton Paris in *The Life of Sir Humphry Davy*, page 406.

CHAPTER TWO: CHEMISTRY

1. Faraday's early observation on the French character can be found in Bowers and Symons's book *Curiosity Perfectly Satisfied: Faraday's Travels in Europe*, page 15.

2. Faraday's slightly later observation of the French character is from his August 6, 1814, letter to Benjamin Abbott's brother, Robert Abbott, given in James (1991–2011), vol. 1, page 80.

3. Faraday expressed this opinion of Lady Davy in a letter on January 25, 1815, to Benjamin Abbot, given in Williams (1965), page 40.

4. Faraday wrote to Benjamin Abbott about the French chemists on February 23, 1815. The letter is given in James (1991–2011), vol. 1, page 128.

5. Faraday commented on the Italian people in an August 6, 1814, letter to Robert Abbott, given in James (1991–2011), vol. 1, page 81.

6. Davy's impressions of Volta are quoted by John Ayrton Paris in *The Life of Sir Humphry Davy*.

7. The incident at the Marcets' dinner party is related by Alan Hirshfeld in *The Electric Life of Michael Faraday*, page 53. Hirshfeld cites as his source Bowers and Symons's *Curiosity Perfectly Satisfied: Faraday's Travels in Europe*.

8. The quotation "We loved Faraday" is from the booklet by M. Dumas (secretary of the Institut Impérial de France), *Éloge historique de Michel Faraday*.

9. Faraday's journal entry about Bonaparte was on March 7, 1815, and is given in Bence Jones, vol. 1, page 115.

10. Faraday wrote this letter to his mother on April 16, 1815. It is given in James (1991–2011) vol. 1, page 128.

11. Faraday's early observations to Benjamin Abbott on the art of lecturing were in a June 13, 1813, letter, given in James (1991–2011), vol. 1, pages 60–63.

12. Berzelius's publication of his censure of Davy is reported in Williams (1965), page 45.

CHAPTER THREE: HISTORY

1. William Gilbert made his comments about "miracle-mongering" in his book *De Magnete*, page 77.

2. The torsion balance was a kind of torsion pendulum, constructed so that electric or magnetic forces of attraction or repulsion could be measured by the twisting of the wire that held the pendulum bob.

3. Newton wrote this letter in 1692. Richard Bentley was a classical scholar

and theologian who, a few years later, became master of Trinity College, Cambridge. It is given in W. D. Niven's *The Scientific Papers of James Clerk Maxwell*, vol. 2, page 316.

4. Ampère's explanation of why it took so long to discover the magnetic effect of an electric current is published Louis de Launay's *Corréspondance du Grand Ampère*, vol. 2, page 556.

5. Oersted wrote that the electric conflict performs circles in his paper "Experiments on the Effect of a Current of Electricity on the Magnetic Needle," published in *Annals of Philosophy*, vol. 16, pages 237–277.

CHAPTER FOUR: A CIRCULAR FORCE

1. The poem ending "'tis love" is on page 73 in vol. 1 of Faraday's *Common Place Book*, held by the Institution of Engineering and Technology, London, and is quoted in Williams (1965), pages 96–97.

2. Faraday wrote this letter to Sarah on July 5, 1820. It is given by Bence Jones, vol. 1, page 317.

3. Faraday's journal entry for this day can be found in Bence Jones, vol. 1, pages 319–20.

4. Faraday wrote to Sarah on August 14, 1863, describing her as "a pillow to my mind," as reported in Frank A. J. L. James's *Michael Faraday: A Very Short Introduction*, page 15.

5. Faraday wrote this letter to Ampère on September 3, 1822. It is given in James (1991–2011), vol. 1, pages 287–88.

6. George's recollection of this moment is reported in Bence Jones, vol. 1, page 345.

7. This journal entry is reported in Gooding and James's book *Faraday Rediscovered*, page 120.

8. Faraday included this comment in one of several letters on the theme of lecturing to Benjamin Abbott in June 1813. They are given in James (1991–2011), vol. 1, pages 55–65.

9. Faraday told Richard Phillips about his "nervous headaches" in a letter of August 29 1828, held in the Preussische Staatbibliotek, Berlin. The letter is quoted by Williams (1865), page 102.

10. Tyndall's summary of Sergeant Anderson's character, and Abbott's account of the boiler incident, are reported in Thompson, pages 96–97.

11. Faraday described his skepticism of theories in a letter to Ampère on February 2, 1822, given in Williams (1965), page 168.

12. Faraday's letter to Ampère, contrasting their working lives, was written on November 17, 1825, and is given in James (1991–2011), vol. 1, page 392.

13. Faraday's letter to Davy's biographer is given in James (1991–2011), vol. 2, page 497.

CHAPTER FIVE: INDUCTION

1. The passage quoted is toward the end of a long letter from Faraday to Richard Phillips. The complete letter may be found in Thompson, pages 109–110.

2. Faraday published these thoughts on the electrotonic state in series 1 of his *Researches*, starting at paragraph 60.

3. Faraday used the expression "magnetic curves" in series 1 of his *Researches*, paragraph 114, but offered the alternative expression "lines of magnetic forces" in a footnote. The expression "lines of magnetic force" (suggesting a physical presence) then came to supersede the purely geometric "magnetic curves" in Faraday's writings.

4. Faraday expressed annoyance about the false accusation of plagiarism in a March 31, 1832, letter to William Jerden, given in James (1991–2001), vol. 2, page 29.

5. Faraday made his apology for "egotism" at the end of a long letter to Richard Phillips, mentioning, among other things, the electrotonic state. The complete letter may be found in Thompson, pages 114–17.

6. Faraday wrote his sealed note to the Royal Society on March 12, 1832. Its text can be found in James (1991–2011), vol. 2, letter 557.

7. Faraday expressed his agnosticism as to what a current actually was in series 3 of his *Researches*, paragraph 283.

8. Faraday summarized his theory of chemically equivalent weights in series 7 of his *Researches*, paragraph 869.

9. Helmholtz's words are quoted in *Twentieth Century Physics*, edited by L. Brown, B. Pippardi, and A. Pais, page 52.

10. The modern explanation of this process is in terms of electrons, which were not discovered for another sixty years. Each negatively charged ion arriving at the anode gives up one or more electrons and each positively charged ion arriving at the cathode receives one or more electrons.

11. John Tyndall described Faraday as "working at the boundaries of knowledge" in his book *Faraday as a Discoverer*, page. 73.

12. Faraday told Whewell about the rough reception the new terms had initially received in a letter on May 15, 1834, given in James (1991–2011), vol. 2, page 186.

13. The reference to Faraday's being "dimly aware" of the lateral repulsion between lines of force can be found in Thompson, page 165.

14. Electric lines of force always run from one charged body to another oppositely charged. So when two similarly charged bodies act on one another, the lines do not run between them but bend around, thus pushing away from one another. This sideways repulsion between the two sets of lines has the appearance of a direct repulsion between the two bodies. A similar sideways repulsion between magnetic lines of force explains why like magnetic poles repel one another.

15. Faraday gave this summary of his theory of the nature of electricity in series 14 of his *Researches*, paragraphs 1,669–1,678.

16. This passage is from Thompson, page 221. It follows the author's account of Faraday's unsuccessful attempt to detect the effect later discovered by Peter Zeeman, also described in our chapter 7.

CHAPTER SIX: A SHADOW OF A SPECULATION

1. Faraday's rueful comment on his loss of memory is reported in Bence Jones, vol. 2, page 142.

2. This comment, written on September 18, 1845, can be found in *Faraday's Diary*, vol. 4, page 227.

3. The giant electromagnet is described by Frank A. J. L. James in his book *Michael Faraday, a Very Short Introduction*, page 80.

4. Faraday presented this image of a man suspended in a magnetic field in series 20 of his *Researches*, paragraph 2,281.

5. The quoted words, including the earlier remark that the aether would have to be "destitute of gravitation but infinite in elasticity," are from Faraday's letter to Richard Phillips on April 15, 1846, published the following month in the *Philosophical Magazine* with the title "Thoughts on Ray-vibrations" and reprinted in series 29 of his *Researches*.

6. The words quoted here are from the same letter to Richard Phillips, which also served as Faraday's article for the *Philosophical Magazine*, as cited above.

7. Tyndall's opinion of Faraday's "Ray-vibrations" is quoted by Thompson, page 193.

8. A fuller version of Faraday's letter to Oersted can be found in Bence Jones, vol. 2, page 268.

9. Faraday commented on his lateness in finding the magnetic effect of flames in an article in the *Philosophical Magazine*, 1847, vol. 31, page 401.

10. Faraday made this comment about space and matter in series 25 of his *Researches*, paragraph 2,787.

11. The passage quoted is from a letter Airy wrote to the Royal Institution's

secretary, the Reverend John Barlow, in February 1855. The complete letter can be found in Bence Jones, vol. 2, page 353.

12. Faraday opened series 29 of his *Researches*, paragraph 3,070, with this reference to magnetic lines of force.

13. Faraday gave this description of what later became known as a tube of magnetic flux in paragraph 3,072 of series 29 of his *Researches*.

14. Faraday stated his law of electromagnetic induction in series 28 of his *Researches*, paragraph 3,115.

15. The right-hand rule: Point the thumb, first finger, and second finger of the right hand at right angles to each other. If the first finger represents the direction of the magnetic field and the thumb the direction of motion of the conductor, then the second finger will represent the direction of the electromotive force generated. A similar rule, but for the left hand, applies to motors. These rules were popularized by Ambrose Fleming.

16. Faraday carried out another experiment to demonstrate that lines of force ran all the way through a permanent magnet—reported in series 29 of his *Researches*, starting at paragraph 3,084.

17. Faraday wrote of the electrotonic state forcing itself on his mind in series 29 of his *Researches*, paragraph 3,269.

18. The third and last volume of Faraday's *Researches* covered series 19 to series 29.

19. Faraday made these observations on table turners in a letter to his friend Christian Friedrich Schönbein. A fuller version can be found in Thompson, page 252.

20. Faraday's letter expressing weariness with table turners is reported in Bence Jones, vol. 2, page 468.

21. Faraday's part-explanation for his failure to take on a pupil is in one of the miscellaneous notes found after his death. A fuller version can be found in Thompson, page 243.

22. This is the introduction to Maxwell's paper "On Faraday's Lines of Force," *Transactions of the Cambridge Philosophical Society*, vol. 10, part 1. (Read on December 10, 1855, and February, 11, 1856).

23. Faraday's reply to Maxwell's first letter is given in Campbell and Garnett, page 252.

24. A complete copy of this November 9, 1857, letter from Maxwell to Faraday can be found in Campbell and Garnett's *The Life of James Clerk Maxwell*, 2nd ed. (London: Macmillan, 1884), page 203.

25. A complete copy of Faraday's November 13, 1857, reply to Maxwell can be found in Campbell and Garnett's *The Life of James Clerk Maxwell*, 2nd ed. (London: Macmillan, 1884), pages 205–206.

CHAPTER SEVEN: FARADAY'S LAST YEARS

1. A fuller version of Faraday's report of this visit to the South Foreland Lighthouse can be found in Williams (1965), page 491.

2. Faraday made this comment on mariners' trust of lighthouses in a report on a proposal by Joseph Watson in 1854 for the use of a form of carbon-arc lamp in lighthouses. Faraday thought Watson's scheme was expensive and impracticable, and it was not adopted.

3. One of the leading advocates *for* the use of poison gas against the Russians in the Crimean War was the secretary of the Department of Science, Lyon Playfair, after whom the Playfair cipher was misleadingly named. The true inventor of this cipher was Charles Wheatstone, whose late defection from a Friday Evening Discourse at the Royal Institution led to Faraday's impromptu "Ray-vibrations" talk. See also note 1 to the introduction.

4. The extract is from evidence Faraday gave to the public-school commissioners in 1862. A fuller account of this episode can be found in Hamilton, pages 388–391.

5. Faraday wrote of his belief in a relation between gravity and electricity in his *Researches*, series 24, paragraph 2,717.

6. A shot tower is a tower built for the production of shot balls by free fall of molten lead.

7. Faraday recorded this final experiment in his laboratory notebook on March 12, 1862.

8. Faraday made this comment on honors to the Chancellor of the Exchequer, Thomas Spring Rice, in 1838. Thompson refers to the matter on page 271.

9. Faraday's comment about not wanting to be "a Sir" is quoted by Thompson, pages 273–74.

10. Instituted by Napoleon, this prize was formally designated the Volta Prize. See note 11 to chapter 1.

11. Faraday's comment about the Prussian knighthood is from the same letter in which he said he did not want to be "a Sir," quoted by Thompson, pages 273–74.

12. Perhaps at a loss how to respond, Schönbein did not reply to Faraday's last letter.

13. Faraday mentioned that he wanted a simple funeral in a January 1866 letter to astronomer Sir James South, given in Bence Jones, vol. 2, page 478.

14. John Tyndall gives this description of Faraday in chapter 4 of his book *Faraday as a Discoverer*, page 37.

15. The extract is from Helmholtz's Faraday lecture to Fellows of the Chemical Society of London on April 5, 1881, quoted by Thompson, pages 282–83.

CHAPTER EIGHT: WHAT'S THE GO O' THAT?

* Quoted passages in this chapter for which the source is not evident from the text or given a numbered note marker are from Campbell and Garnett.

1. The quoted line is from Robert Burns's poem "To a Mouse," published in *Poems, Chiefly in the Scottish Dialect* (Kilmarnock: John Wilson, 1786).

2. David Forfar gives an excellent brief history of the Clerk and Cay families in his article "Generations of Genius."

3. Maxwell's question "What's the go o' that?" is reported by Campbell and Garnett, page 12.

4. A complete version of Maxwell's "Song of the Edinburgh Academician" can be found in Campbell and Garnett, pages 292–293.

CHAPTER NINE: SOCIETY AND DRILL

* Quoted passages in this chapter for which the source is not evident from the text or given a numbered note marker are from Campbell and Garnett.

1. The fellow student who reported on Maxwell's midnight jogging was Charles Hope Robertson. Robertson was also the friend whom Maxwell helped by reading out the next day's bookwork when he had eye trouble.

2. The complete essay "Are There Any Real Analogies in Nature?" can be found in Campbell and Garnett, pages 235–244.

3. The fellow student who "never met a man like" Maxwell was W. N. Lawson.

4. William Hopkins was the most successful Cambridge coach of the time. Seventeen of his pupils became senior wranglers, including George Gabriel Stokes, William Thomson, P. G. Tait, Maxwell's rival E. J. Routh, and Arthur Cayley, who created the theory of matrices.

5. For the Lagrangian and the Hamiltonian, see note 5 to chapter 13. The Routhian combines elements of the Lagrangian and the Hamiltonian, and the Laplacian is, in vector terminology, the divergence of the gradient of a scalar function.

6. It was W. N. Lawson who reported to Campbell on Maxwell's geniality and kindness.

CHAPTER TEN: AN IMAGINARY FLUID

1. Young Maxwell's question about the blue stone is reported in Campbell and Garnett, page 14.

2. Maxwell's reference to "different kinds of paint" can be found in W. D. Niven's *The Scientific Papers of James Clerk Maxwell*, vol. 1, page 127.

3. An instance of Maxwell's enduring influence is the present-day chromaticity diagram, which differs only in detail from his original color triangle. It uses a right-angled triangle in which the proportions of red and green are plotted, the proportion of blue being implied because the proportions of red, green, and blue always add up to 1.

4. Maxwell's letters to Thomson at this time are in Harman, vol. 1, pages 254–63 and 319–20.

5. The full text of Maxwell's comments on Faraday's experimental methods (which he contrasts with those of Ampère) can be found in article 528 of his *Treatise on Electricity and Magnetism*, on page 176 of vol. 2.

6. The quoted words from Maxwell's paper "On Faraday's Lines of Force," can be found in Simpson, page 57.

7. The quoted words from Maxwell's paper "On Faraday's Lines of Force," can be found in Simpson, page 60.

8. Airy's disparaging comment about lines of force is quoted by J. J. Thomson in his essay "James Clerk Maxwell" in *James Clerk Maxwell: A Commemoration Volume*, page 28.

9. The flux concentration, or density, in a small region (or, in the limit, at a point) is the amount of flux per unit area of a plane surface perpendicular to the direction of the flux.

10. Thomas K. Simpson gives a guided study of part 1 of Maxwell's paper "On Faraday's Lines of Force" in his book *Maxwell on the Electromagnetic Field*.

11. Maxwell wrote of the power of the subconscious mind in a May 29, 1857, letter to his friend Litchfield. It is given in Campbell and Garnett, page 136.

12. These lines are from Maxwell's poem "Recollections of Dreamland," written a few months after his father died. The complete poem can be found in Campbell and Garnett, pages 298–99.

13. Maxwell's preference for "the rubs of the world" is reported in Campbell and Garnett, page 126.

14. The rival candidate for whom Maxwell provided a reference was William Swan, later professor of natural philosophy at St. Andrews.

15. These lines are also from "Recollections of Dreamland," mentioned above in note 12.

16. Maxwell made this reference to the "Natural Philosophers of the North" in an October 14, 1856, letter to Cecil Monro, quoted in Campbell and Garnett, page 132.

CHAPTER ELEVEN: NO JOKES ARE UNDERSTOOD HERE

*Quoted passages in this chapter for which the source is not evident from the text or given a numbered note marker are from Campbell and Garnett.

1. Maxwell's comment "No jokes are understood here" is report by Ivan Tolstoy in his book *James Clerk Maxwell*, page 80.

2. This passage and the following one, from Maxwell's inaugural address at Marischal College, are given by R. V. Jones in his essay "The Complete Physicist: James Clerk Maxwell, 1831–79," *The Yearbook of the Royal Society of Edinburgh*, 1980.

3. The quoted words from Maxwell's paper "On Faraday's Lines of Force" can be found in Simpson, page 57.

4. Faraday wrote this letter to Maxwell on November 13, 1857. It is given in Campbell and Garnett, page 145.

5. This is the second of four verses of Maxwell's poem "The Song of the Atlantic Telegraph Company." The complete poem can be found in Campbell and Garnett, page 140.

6. These are the last four of eight verses of Maxwell's poem "To K. M. D." The complete poem can be found in Campbell and Garnett, pages 302–303.

7. Maxwell put the question "If you go at 17 miles a minute . . . ?" in a letter to P. G. Tait that is now in the Cambridge University Archive.

8. To statisticians, the Maxwell distribution is the square root of a chi-square distribution with three degrees of freedom.

9. This account of Maxwell as his students at Aberdeen saw him is from George Reith, who became moderator of the Church of Scotland and father of Lord Reith, the first governor of the British Broadcasting Corporation. George Reith's comments are reported by R. V. Jones in his essay "The Complete Physicist: James Clerk Maxwell, 1831–79" in the *Yearbook of the Royal Society of Edinburgh*, 1980.

10. David Gill reminisced about Maxwell's classes in the introductory part of his book *History and Description of the Royal Observatory Cape of Good Hope*, pages xx–xxi.

11. This farmer's recollection is reported by R. V. Jones in his essay "The Complete Physicist: James Clerk Maxwell, 1831–79" in the *Yearbook of the Royal Society of Edinburgh*, 1980.

CHAPTER TWELVE: THE SPEED OF LIGHT

1. Maxwell's inaugural lecture at King's College, London, is given in Harman, vol. 1, pages 662–74.

2. Amazingly, no one managed to repeat the feat, and many years passed before the next color photograph appeared. About one hundred years later, a team at Kodak Research Laboratories discovered that the photographic plates Maxwell and Sutton had used were completely insensitive to red light! By coincidence, the plates *were* sensitive to some ultraviolet light and the solution the experimenters had used as a red filter happened to have a passband in just the right part of the ultraviolet spectrum. Ultraviolet had acted as a surrogate for red. Lucky Maxwell! But perhaps he made his own luck. It was a rule with him never to discourage a man from doing an experiment, however slim the apparent prospect of success. Arthur Schuster, one of Maxwell's students at the Cavendish, recalled Maxwell saying of another student, "if he doesn't find what he is looking for, he may find something else."

3. For clarity, the laws are expressed here in a modern way; in 1861, the terms *field* and *flux* were not yet in general use. Law 3 is now often called Ampère's Law and law 4 is called Faraday's Law.

4. Maxwell's medium of cells and idle wheels, like the incompressible fluid in his model from his Cambridge days, had an built-in inverse-square law. This followed from the property of magnetic flux (as modeled by the spinning of the cells) that the amount of flux entering a region enclosed by any closed surface was the same as the amount leaving. The reasoning, though more mathematical, is essentially the same as that given in chapter 10 for Maxwell's fluid model, where the amount of fluid emerging from any sphere with a point source at its center is the same, no matter what the size of the sphere. More generally, any effect that spreads out uniformly in three-dimensional space follows an inverse-square law. The principle is embodied is the divergence theorem, otherwise known as Gauss's theorem, in vector analysis.

5. Maxwell identified Faraday's electrotonic state with the magnetic vector potential (a quantity whose curl at any point in the field was equal to the magnetic flux density there). Magnetic flux, represented by the rotation of the cells, was, in Maxwell's interpretation, the electromagnetic momentum of the field.

6. Thomas K. Simpson gives a guided study of Maxwell's paper "On Physical Lines of Force" in his book *Maxwell on the Electromagnetic Field.*

7. This passage from Maxwell's paper "On Physical Lines of Force" is quoted by Richard Glazebrook in his book *James Clerk Maxwell and Modern Physics*, page 173.

8. See note 4. The same reasoning about the inverse-square law applies, this time to electric flux, as modeled by the distortion of the cells.

9. The ratio is the number of electrostatic units in one electromagnetic unit—broadly, the relative strengths of the two types of force. It had the dimensions of velocity because electromagnetic force depends not only on the quantity of charge but also on its velocity. The electromagnetic unit is bigger because it takes more charge to generate a given force by magnetic action when the charge is moving at unit velocity than it does by electrostatic action.

10. The quoted words from Maxwell's paper "On Physical Lines of Force" can be found in Simpson, page 216.

11. Monro's letter to Maxwell about the "brilliant result" is given in Campbell and Garnett, page 163.

12. Newton made these comments in the letter to Richard Bentley quoted in chapter 3—see note 3 to that chapter. The failure of Newton's followers to heed this warning may be due in part to the misplaced zeal of his disciple and evangelist Roger Coates, who wrote in a preface to Newton's *Principia Mathematica* that action at a distance is one of the primary properties of matter.

13. Faraday's skepticism of atoms is reported by Simon Blackburn in his book *Think: A Compelling Introduction to Philosophy*, page 248.

14. Charles Coulson was professor of theoretical physics at King's College, London, from 1947 to 1952. One of his doctoral students there was Peter Higgs, who later created the theory of the Higgs field and its associated particle, the Higgs Boson.

CHAPTER THIRTEEN: GREAT GUNS

1. Maxwell's "great guns" letter to his cousin was written in January 1865. The complete letter can be found in Campbell and Garnett, pages 168–69.

2. These words from Maxwell's paper "A Dynamical Theory of the Electromagnetic Field" can be found in Simpson, page 255.

3. These words from Maxwell's paper "A Dynamical Theory of the Electromagnetic Field" can be also found in Simpson, page 255.

4. Thomson and Tait published their *Treatise on Natural Philosophy* in 1867 after seven years of collaborative effort, and after a difficult gestation it sold very well. By the usual alphabetical convention, Tait should have been the first-named author, but Thomson's stellar reputation took precedence. It was Tait, however, who did most of the work: he was forever chasing Thomson to comment on drafts he'd sent. Thomson responded to Tait's scolding with good humor and they remained lifelong friends.

5. Lagrange's set of equations embodied the so-called principle of least action,

first formulated by Pierre Louis Moreau de Maupertuis in 1746, and its characteristic function (the difference between the physical system's kinetic energy and its potential energy) became known as the Lagrangian. William Rowan Hamilton extended Lagrange's method to form an alternative system of equations now almost universally used to describe the dynamics of systems. His characteristic function, the Hamiltonian, represents the total energy of a system.

6. Maxwell included the "belfry" passage in a review for *Nature* of the second edition of Thomson and Tait's *Treatise on Natural Philosophy* in 1879. He was making the point that it might never be possible to give an explanatory model of electrodynamics, and he had probably also used the belfry analogy in his lectures to students.

7. A guided study of Maxwell's paper "A Dynamical Theory of the Electromagnetic Field" is given by Thomas K. Simpson in his book *Maxwell on the Electromagnetic Field.*

8. Maxwell explained why he did not condense his equations in his *Treatise on Electricity and Magnetism*, vol. 2, page 254.

9. Thomson's view that Maxwell had "lapsed into mysticism" is reported by Templeton and Herrmann in their book *The God Who Would Be Known: Revelations of Divine Contemporary Science*, page 161.

10. Gill made these comments about Maxwell's teaching in the introductory part of his book *History and Description of the Royal Observatory Cape of Good Hope*, pages xx–xi.

CHAPTER FOURTEEN: COUNTRY LIFE

1. This is the first of four verses of Maxwell's poem "Will You Come along with Me?" The complete poem can be found in Campbell and Garnett, page 301.

2. This and the following passage are quoted by Campbell and Garnett on page 180, though they do not identify the observer.

3. These passages from Maxwell's essay "Is Autobiography Possible?" are given in Campbell and Garnett, page 125.

4. The quotation, given by Campbell and Garnett on page 196, is from a draft found after Maxwell's death. The draft ends at ". . . continually," but the final "changing" is clearly implied.

5. Maxwell made this comment about his decision not to apply for the St. Andrews post in an October 30, 1868, letter to William Thomson. The letter is held in the Glasgow University Library.

6. Maxwell's letter seeking advice on political matters was to W. R. Grove,

who was vice president of the Royal Institution. The complete letter can be found in Harman, vol. 2, page 461. Grove was not only a scientist but also a successful barrister who became a judge of the Queen's Bench. As far as he was political at all, Maxwell was a Conservative, and the Conservatives lost the election held shortly before the appointment was made.

CHAPTER FIFTEEN: THE CAVENDISH

*Quoted passages in this chapter for which the source is not evident from the text or given a numbered note marker are from Campbell and Garnett.

1. Maxwell spoke of Alexander Graham Bell's father when giving the Rede lecture at Cambridge in 1878. His title for the lecture was "On the Telephone"; a longer extract from it can be found in Campbell and Garnett, pages 177–78.

2. Examples may be found in Campbell and Garnett on pages 16 and 18.

3. Maxwell gave this premonition of chaos theory in 1873 in an essay for the Eranus Club: "Does the Progress of Physical Science Give Any Advantage to the Opinion of Necessity (or Determinism) over That of the Contingency of Events and the Freedom of the Will?" For a wider audience he would no doubt have shortened the title to something like "Science and Free Will." The complete essay can be found in Campbell and Garnett, pages 209 to 213.

4. The current owner of Glenlair, a great admirer of Maxwell, has gone to great lengths to have the remains of the house made safe for visitors, and has established a visitor center in the original porch. At the time of this writing, the house was temporarily closed for further restoration.

5. Heaviside did not include this eulogy of Maxwell in any of his publications, but one of his American followers, Ernst J. Berg, recorded it in "Oliver Heaviside, a Sketch of His Work and Some Reminiscences of His Later Years," published in the *Journal of the American Academy of Sciences* in 1930, and a fuller version can also be found in Rollo Appleyard's *Pioneers of Electrical Communication*, page 257.

CHAPTER SIXTEEN: THE MAXWELLIANS

Our search for Einstein's statement "One scientific epoch ended and another began with James Clerk Maxwell" has been unfruitful, but the quote can be found, for example, on the websites of King's College, London, and the National High Magnetic Field Laboratory, Florida State University.

Feynman's observation on the historical importance of Maxwell's discovery of the laws of electrodynamics is from *The Feynman Lectures on Physics*, by Richard Feynman, Robert Leighton, and Matthew Sands, vol. 2, chap. 1, p. 11 (1964).

1. Heaviside wrote of his first sight of Maxwell's *Treatise on Electricity and Magnetism* to a French admirer, Joseph Bethenod, in 1918, and Bethenod included it (translated into French) in an obituary to Heaviside published in *Annales des postes télégraphes et téléphones* in 1925, pages 521 to 538. The original letter has not survived, and we are greatly indebted to Paul J. Nahin for translating the French version back into English and including it in his book *Oliver Heaviside: Sage in Solitude*. Both Bethenod and Nahin were careful to translate as literally as possible, so we can be confident that the words are close to Heaviside's own.

2. Heaviside compared himself to "Old Teufelsdröckh" in a July 1908 letter to Joseph Larmor. Diogenes Teufelsdröckh was Thomas Carlyle's alter ego in his satirical book *Sartor Resartus*. Teufelsdröckh was an eccentric professor whose "philosophy of clothes" provided metaphors for the author's own thoughts on life.

3. Heaviside made this reference to the sea serpent in his article "The Earth as a Return Conductor," first published in the *Electrician* and reprinted in as article 23 of his *Electrical Papers*, vol. 1, pages 190–95.

4. Fitzgerald wrote of Heaviside's achievement in condensing Maxwell's theory in an 1893 *Electrician* review of Heaviside's collected *Electrical Papers*.

5. The (scalar) electric or magnetic potential at any point in the field is the energy needed to move a unit electric charge or a unit magnetic pole there from an infinite distance away. Maxwell used fluid pressure as an analogy for these potentials in his fluid model. There is also the magnetic vector potential that Maxwell identified with Faraday's electrotonic state; its curl gives the magnetic flux density at any point in the field.

6. Heaviside explained his view of the potentials in an article that can be found in his *Electrical Papers*, vol. 2, pages 483–85.

7. μ and ε are the ratio of the electric and magnetic flux densities to their respective field forces. In empty space they are usually given the subscript 0, omitted here for simplicity.

8. In the presence of electric charge density ρ and current density $\mathbf{J}$, the equations become:

$$\text{div } \mathbf{E} = \rho/\varepsilon$$
$$\text{div } \mathbf{H} = 0$$
$$\text{curl } \mathbf{E} = -\mu\partial\mathbf{H}/\partial t$$
$$\text{curl } \mathbf{H} = \varepsilon\partial\mathbf{E}/\partial t + \mathbf{J}$$

The electric flux density $\mathbf{D}$ (= $\varepsilon\mathbf{E}$)—Maxwell called it displacement—may be used rather than the electric field intensity vector $\mathbf{E}$, and today the magnetic flux density

B ($= \mu$**H**) is now generally used rather than the magnetic field intensity vector **H**. The constants ε and μ depend on the medium. If the medium is nonisotropic, they are replaced by tensors and the equations need to be adjusted accordingly.

9. Maxwell himself described the third equation, curl $\mathbf{E} = -\mu \partial \mathbf{H}/\partial t$, though not using these symbols, in a "Note on the Electromagnetic Theory of Light" in 1868. This was a note added to a report on his experiment with Charles Hockin to determine the ratio of the electromagnetic and electrostatic units of charge. Maxwell did not include the equation in his *Treatise*; nor, it seems, did he mention it to his students, and it stayed in the dark until his collected papers were published in two volumes in *The Scientific Papers of James Clerk Maxwell*, edited by W. D. Niven in 1890, five years after Heaviside had published the four now-famous equations.

10. Heaviside published his reformulation of Maxwell's theory in section 4 of a long series of papers, "Electromagnetic Induction and Its Propagation," which appeared in installments in the weekly journal the *Electrician*, starting in 1885. The first half of this series (including section 4) was reprinted in vol. 1 of his collected *Electrical Papers* as article 30, and the second half in vol. 2 as article 35.

11. In vector algebra the vector product of two vectors is itself a vector. Its magnitude is the arithmetic product of those of the two vectors multiplied by the sine of the angle between them, and its direction is at right angles to both the two vectors.

12. Heaviside also gave the formula for energy flow in Section 4 of the series of papers "Electromagnetic Induction and its Propagation," reprinted as Article 30 in Vol. 1 of his collected *Electrical Papers*.

13. Lodge decided to edit out the words *eccentric* and *repellent* when he published his collected papers on lightning protection in 1892. (He had by then become a good friend of Heaviside.)

14. The French physicist and philosopher Pierre Duhem gave this description of Lodge's book as an example of what he saw as British physicists' over-reliance on physical models in his own book, which has been translated as *The Aim and Structure of Physical Theory*, pages 70–71.

15. Heaviside wrote this tribute to Fitzgerald in a January 1901 letter. Bruce Hunt quotes it in *The Maxwellians*, page 187.

16. Fitzgerald wrote of this "great difficulty" to J. J. Thomson on December 23, 1884. The letter is quoted by Bruce Hunt in *The Maxwellians*, page 45.

17. Hermann Helmholtz was ennobled (acquired the *von*) in 1882.

18. This passage is from a letter Hertz wrote to Heaviside on March 21, 1889, quoted by Rollo Appleyard in his book *Pioneers of Electrical Communication*, page 238. In the same letter, he emphatically endorsed Heaviside's abandonment of the electric and magnetic potentials.

19. Heaviside made this comment about Helmholtz's theory being "Maxwell's

run mad" in a June 15, 1892, letter to Lodge, quoted by Bruce Hunt in *The Maxwellians*, page 198.

20. Heaviside congratulated Hertz on giving a death blow to "these theories" (those employing action at a distance) in a July 13, 1889, letter, quoted by Paul Nahin in *Oliver Heaviside: Sage in Solitude*, page 111. Had Heaviside known how much Hertz revered Helmholtz (whose own theory had action-at-a-distance elements), he may not have rubbed salt in the wound so forcefully. But with Heaviside one cannot be sure.

CHAPTER SEVENTEEN: A NEW EPOCH

1. Tait called Heaviside's and Gibb's vector analysis "a hermaphrodite monster" in the preface to the third (1890) edition of his book *An Elementary Treatise on Quaternions*. Heaviside's response, and other comments mocking Tait's devotion to quaternions, are recorded in vol. 1, chapter 3 of his treatise *Electromagnetic Theory*. Tait would have seen them when they appeared earlier in journals, mostly the *Electrician*.

2. The coherer was a kind of switch that was activated by electromagnetic radiation—a tubeful of metallic powder that was ordinarily a very poor conductor of electricity but became a very good one when the magnetic effect of radiation caused its particles to cohere and thus provide a low-resistance path.

3. For many years nobody knew for certain how Marconi's transatlantic signals managed to follow the curvature of Earth. Oliver Heaviside and Arthur Kennelly, an expatriate Briton living in America, independently postulated an ionized layer in the upper atmosphere that reflected the waves. Experiments by Edward Victor Appleton and Miles Barnett in the 1920s confirmed that they were correct.

4. Hertz made this much-quoted observation in his 1892 book, translated as *Electric Waves: Being Researches on the Propagation of Electric Action with Finite Velocity through Space*, page 21.

5. The quotation is from "Maxwell's Influence on the Development of the Conception of Physical Reality," an essay by Albert Einstein in *James Clerk Maxwell, A Commemorative Volume*, published in 1931.

6. Planck wrote of his "act of desperation" to Professor Robert Williams Wood of Johns Hopkins University, Baltimore, on October 7, 1931. The letter is given and discussed by Malcolm Longair in his book *Theoretical Concepts in Physics*, pages 222–23.

7. Feynman's comment on the importance of potentials in quantum electrodynamics is recorded in *The Feynman Lectures on Physics*, by Richard Feynman, Robert Leighton, and Matthew Sands, vol. 2, chapter 1, p. 3 (1964).

8. In Michelson and Morley's experiment, each part of the light beam was reflected back toward the point where the beam was split and the experiment was designed to detect a difference in their average speeds, back and forth. The average speed of part of the beam traveling closer to the direction of the aether drift was expected to be slightly slower. The reason for this can be seen by considering the components of each part of the beam along the line of the aether drift and transverse to it. Light traveling back and forth along the line of aether drift would lose more time traveling upstream than it would gain traveling downstream, and although light traveling transverse to the aether drift would also be slowed (because, relative to the aether, it was covering a greater distance), this slowing effect could be shown mathematically to be less than that on light traveling back and forth along the line of aether drift.

9. Poincaré first made this statement on the nonexistence of absolute motion and time in his paper "La théorie de Lorenz et le principe de réaction," published in 1900, available in *Archives néerlandaises des sciences exactes et naturelles* 5, pages 252–78.

10. Poincaré also gave this result in 1890 in his paper "La théorie de Lorentz et le principe de Réaction," published in *Archives néerlandaises des sciences exactes et naturelles* 5, series 2, pages 252–78.

11. Einstein published the main part of the special theory of relativity in his 1905 paper "On the Electrodynamics of Moving Bodies" and the derivation of $E = mc^2$ in a short follow-up paper. This paper appeared in the *Annalen der Physik*, vol. 18, pages 639–41 (1905).

12. Einstein's statement is reported by Frederick Seitz in his article about Maxwell in the *Proceedings of the American Philosophical Society* 145, no.1, March 2001.

13. In 1979, Abdus Salam, Sheldon Glashow, and Steven Weinberg gained the Nobel Prize in Physics for showing the electromagnetic force and the weak nuclear force to be different aspects of a single force, now called the electroweak force.

BIBLIOGRAPHY

Appleyard, Rollo. *Pioneers of Electrical Communication*. London: Macmillan, 1930.

Bell, Eric Temple. *Men of Mathematics*. 2 vols. Harmondsworth: Penguin Books, 1965. First published 1937.

Bence Jones, Henry. *The Life and Letters of Faraday*. 2 vols. Philadelphia: Lippincott, 1870.

Blackburn, Simon. *Think: A Compelling Introduction to Philosophy*. Oxford: Oxford University Press, 1991.

Bowers, Brian, and Lenore Symons. *Curiosity Perfectly Satisfied: Faraday's Travels in Europe, 1813–1815*. London: Peter Peregrinus in association with the Science Museum, 1991.

Brown, George Ingham. *Scientist, Soldier, Statesman, Spy: Count Rumford, the Extraordinary Life of a Scientific Genius*. Stroud: Sutton Publishing, 1999.

Brown, L., B. Pippard, and A. Pais, eds. *Twentieth Century Physics*. New York: IOP Publishing, AIP Press, 1995.

Buchwald, Jed Z. *From Maxwell to Microphysics*. Chicago: University of Chicago Press, 1985.

Campbell, Lewis, and William Garnett. *The Life of James Clerk Maxwell*. London: Macmillan, 1882. Second edition published 1884. We have used the online version by Sonnet Software (second edition, 1999), available at www.sonnetsoftware.com/bio/maxbio.pdf, accessed December 9, 2013.

Darrigol, Olivier. *Electrodynamics from Ampère to Einstein*. Oxford: Oxford University Press, 2000.

Davy, John. *Memoirs of the Life of Sir Humphry Davy*. London: Smith Elder, 1836.

de Launay, Louis, ed. *Corréspondance du Grand Ampère*. 3 vols. Paris: Gauthier Villars, 1936–1943.

Duhem, Pierre. *The Aim and Structure of Physical Theory*. Princeton: Princeton University Press, 1954.

Dumas, M. *Éloge historique de Michael Faraday*. Paris: Firmin Didot, 1868.

Dyson, Freeman J. "Why Is Maxwell's Theory So Hard to Understand?" In the *James Clerk Maxwell Commemorative Booklet*. Edinburgh: James Clerk Maxwell Foundation, 1999.

Einstein, Albert. "Maxwell's Influence on the Development of the Conception of

Physical Reality." In *James Clerk Maxwell, A Commemorative Volume*. Cambridge: Cambridge University Press, 1931.

———. *Relativity: The Special and General Theory*. London: Methuen, 1920.

Einstein, Albert, and Leopold Infeld. *The Evolution of Physics*. New York: Simon and Schuster, 1938.

Everitt, C. W. Francis. *James Clerk Maxwell: Physicist and Natural Philosopher*. New York: Charles Scribner's Sons, 1975.

———. "Maxwell's Scientific Creativity." In *Springs of Scientific Creativity*, edited by Rutherford Aris, H. Ted David, and Roger Stuewer. Minneapolis: University of Minnesota Press, 1983.

———. "Maxwell's Scientific Papers." *Applied Optics* 6, no. 4 (1967).

Faraday, Michael. *Common Place Book*. London: Institution of Electrical Engineers.

———. *Experimental Researches in Electricity*. New York: Dover Publications, 1965. Originally published in the *Philosophical Transactions of the Royal Society*, 1831–1852.

———. *Faraday's Diary, Being the Various Philosophical Notes of Experimental Investigation*. Edited by Thomas Martin. London: Bell and Sons, 1932–1936.

Feynman, Richard P., Robert B. Leighton, and Matthew Sands. *Lectures on Physics*. New York: Addison Wesley, 1965.

Fleisch, Daniel. *A Student's Guide to Maxwell's Equations*. Cambridge: Cambridge University Press, 2008.

Fleming, Ambrose. "Some Memories." In *James Clerk Maxwell, A Commemorative Volume*. Cambridge: Cambridge University Press, 1931.

Forfar, David O. "Generations of Genius." In the *James Clerk Maxwell Commemorative Booklet*. Edinburgh: James Clerk Maxwell Foundation, 1999.

Forfar, David O., and Chris Prichard. "The Remarkable Story of Maxwell and Tait." In the *James Clerk Maxwell Commemorative Booklet*. Edinburgh: James Clerk Maxwell Foundation, 1999.

Garnett, William. "Maxwell's Laboratory." In *James Clerk Maxwell, A Commemorative Volume*. Cambridge: Cambridge University Press, 1931.

Gilbert, William. *De Magnete*. New York: Dover Publications, 1958. First published in Latin, 1600. English translation by P. Fleury Mottelay, 1893.

Gill, David. *History and Description of the Royal Observatory Cape of Good Hope*. Edinburgh: Neill, 1913.

Gladstone, John Hall. *Michael Faraday*. London: Macmillan, 1872.

Glazebrook, Richard T. "Early Days of the Cavendish Laboratory." In *James Clerk Maxwell, A Commemorative Volume*. Cambridge: Cambridge University Press, 1931.

———. *James Clerk Maxwell and Modern Physics*. London: Cassell, 1901.

Goldman, Martin. *The Demon in the Aether: The Life of James Clerk Maxwell*. Edinburgh: Paul Harris Publishing, 1983.

Gooding, David, and Frank A. J. L. James. *Faraday Rediscovered: Essays on the Life and Work of Michael Faraday*. New York: American Institute of Physics, 1989.

Hamilton, James. *Faraday: The Life*. London: Harper Collins, 2002.

Harman, Peter M. *Energy, Force and Matter: The Conceptual Development of Nineteenth-Century Physics*. Cambridge: Cambridge University Press, 1982.

———. *The Natural Philosophy of James Clerk Maxwell*. Cambridge: Cambridge University Press, 1998.

———, ed. *The Scientific Papers and Letters of James Clerk Maxwell*. 3 vols. Cambridge: Cambridge University Press, 1990–2002.

Heaviside, Oliver. *Electrical Papers*. 2nd ed. 2 vols. Providence, RI: Chelsea Publishing, 1970. Originally published 1892.

———. *Electromagnetic Theory*. 3 vols. New York: Dover Publications, 1950. First published 1893–1912.

Hertz, Heinrich. *Electric Waves: Being Researches on the Propagation of Electric Action with Finite Velocity through Space*. Translated from the German by Daniel Evan Jones. London: Macmillan, 1893.

Hirshfeld, Alan. *The Electric Life of Michael Faraday*. New York: Walker, 2006.

Hoffmann, Banesh. *The Strange Story of the Quantum*. Harmondsworth: Penguin Books, 1963. First published 1947.

Hunt, Bruce J. *The Maxwellians*. Ithaca, NY: Cornell University Press, 1994.

James, Frank, A. J. L., ed. *The Correspondence of Michael Faraday*. 6 vols. London: Institution of Engineering and Technology, 1991–2011.

———. *Michael Faraday: A Very Short Introduction*. Oxford: Oxford University Press, 2010.

Jeans, James. "James Clerk Maxwell's Method." In *James Clerk Maxwell, A Commemorative Volume*. Cambridge: Cambridge University Press, 1931.

Jones, Reginald Victor. "The Complete Physicist: James Clerk Maxwell, 1831–79." In *Yearbook Royal Society of Edinburgh*, 1980.

Knott, Cargill Gilston. *Life and Scientific Work of Peter Guthrie Tait*. Cambridge: Cambridge University Press, 1911.

Kuhn, Thomas S. *The Structure of Scientific Revolutions*. Chicago: University of Chicago Press, 1962.

Lamb, Horace. "Clerk Maxwell as Lecturer." In *James Clerk Maxwell, A Commemorative Volume*. Cambridge: Cambridge University Press, 1931.

Larmor, Joseph. "The Scientific Environment of James Clerk Maxwell." In *James Clerk Maxwell, A Commemorative Volume*. Cambridge: Cambridge University Press, 1931.

Leff, Harvey S., and Andrew F. Rex. *Maxwell's Demon, Entropy, Information, Computing*. Bristol: Adam Hilger, 1990.

Lindley, David. *Degrees Kelvin: A Tale of Genius, Invention and Tragedy*. Washington, DC: Joseph Henry, 2004.

Lodge, Oliver. "Clerk Maxwell and the Wireless Telegraph." In *James Clerk Maxwell, A Commemorative Volume*. Cambridge: Cambridge University Press, 1931.

Longair, Malcolm S. *Theoretical Concepts in Physics*. Cambridge: Cambridge University Press, 1984.

Mahon, Basil. *The Man Who Changed Everything: The Life of James Clerk Maxwell*. Chichester: Wiley, 2003.

———. *Oliver Heaviside: Maverick Mastermind of Electricity*. London: Institution of Engineering and Technology, 2009.

Marcet, Jane Haldimand. *Conversations on Chemistry*. 9th American ed. Hartford, CT: Cooke, 1824. First published in London, 1806.

Maxwell, James Clerk. "A Dynamical Theory of the Electromagnetic Field." Edited and introduced by Thomas F. Torrance. Edinburgh: Scottish Academic Press, 1982.

———. *Matter and Motion*. Reprinted with notes and appendices by Joseph Larmor, 1920. New York: Dover Publications, 1991.

———. *A Treatise on Electricity and Magnetism*. 3rd ed. Oxford: Clarendon, 1891. Reprinted by Oxford University Press 1998. First edition published 1873.

Nahin, Paul J. *Oliver Heaviside: Sage in Solitude*. New York: IEEE Press, 1987.

Niven, William Davidson, ed. *The Scientific Papers of James Clerk Maxwell*. 2 vols. Cambridge: Cambridge University Press, 1890.

Paris, John Ayrton. *The Life of Sir Humphry Davy*. London: Colburn and Bentley, 1831.

Planck, Max. "Maxwell's Influence on Theoretical Physics in Germany." In *James Clerk Maxwell, A Commemorative Volume*. Cambridge: Cambridge University Press, 1931.

Poincaré, Henri. *La science et l'hypothèse*. Paris: E. Flammarion, 1917. First published 1902.

Pritchard, Chris. "Aspects of the Life and Work of Peter Guthrie Tait." In the *James Clerk Maxwell Commemorative Booklet*. Edinburgh: James Clerk Maxwell Foundation, 1999.

Segrè, Emilio. *From Falling Bodies to Radio Waves: Classical Physicists and Their Discoveries*. New York: W. H. Freeman, 1984.

Siegel, Daniel M. *Innovation in Maxwell's Electromagnetic Theory*. Cambridge: Cambridge University Press, 1991.

Simpson, Thomas K. *Maxwell on the Electromagnetic Field*. New Brunswick, NJ: Rutgers University Press, 1997.

Reid, John S. "James Clerk Maxwell's Scottish Chair." In the *James Clerk Maxwell Commemorative Booklet*. Edinburgh: James Clerk Maxwell Foundation, 1999.

Templeton, John Marks, and Robert L. Herrmann. *The God Who Would Be Known: Revelations of Divine Contemporary Science*. Philadelphia and London: Templeton Foundation, 2002.

Thompson, Sylvanus P. *Michael Faraday, His Life and Work*. London: Cassell, 1901.

Thomson, J. J. "James Clerk Maxwell." In *James Clerk Maxwell, A Commemorative Volume*. Cambridge: Cambridge University Press, 1931.

Tolstoy, Ivan. *James Clerk Maxwell: A Biography*. Edinburgh: Canongate, 1981.

Tyndall, John. *Faraday as a Discoverer*. London: Longmans, Green, 1868.

Watts, Isaac. *The Improvement of the Mind, Also His Posthumous Works*. Edited by Philip Doddridge and David Jennings. London: William Baynes, 1819.

Weaver, Jefferson Hane, and Lloyd Motz. *The Story of Physics*. New York: Avon Books, 1989.

Weightman, Gavin. *Signor Marconi's Magic Box*. London: Harper Collins, 2003.

Whittaker, E. T. *A History of the Theories of Aether and Electricity*. New York: Dover Publications, 1989. First published by Thomas Nelson and Sons, 1951.

Williams, L. Pearce. *Michael Faraday: A Biography*. New York: Basic Books, 1965.

———. *The Origins of Field Theory*. Lanham, MD: University Press of America, 1980.

Yavetz, Ido. *From Obscurity to Enigma: The Work of Oliver Heaviside, 1872–1889*. Basel: Birkhäuser Verlag, 1995.

INDEX